River Life

and the

Upspring of Nature

River Life

and the

Upspring of Nature

NAVEEDA KHAN

DUKE UNIVERSITY PRESS
Durham and London
2023

© 2023 DUKE UNIVERSITY PRESS

All rights reserved

Project Editor: Lisa Lawley

Designed by A. Mattson Gallagher

Typeset in Literata TT by Westchester Publishing Services

Library of Congress Cataloging-in-Publication Data

Names: Khan, Naveeda Ahmed, [date] author.

Title: River life and the upspring of nature / Naveeda Khan.

Description: Durham : Duke University Press, 2023. | Includes
bibliographical references and index.

Identifiers: LCCN 2022029323 (print)

LCCN 2022029324 (ebook)

ISBN 9781478019398 (paperback)

ISBN 9781478016731 (hardcover)

ISBN 9781478024002 (ebook)

Subjects: LCSH: Human ecology—Bangladesh—Jamuna River. |
Rural poor—Bangladesh—Jamuna River. | Sand bars—
Bangladesh—Jamuna River. | River settlements—Bangladesh—
Jamuna River. | BISAC: SOCIAL SCIENCE / Anthropology /
Cultural & Social | NATURE / Ecosystems & Habitats / Rivers

Classification: LCC GF696.B3 K436 2023 (print) | LCC GF696.B3
(ebook) | DDC 304.2086/942095492—dc23/eng20221003

LC record available at https://lccn.loc.gov/2022029323

LC ebook record available at https://lccn.loc.gov/2022029324

Cover art: Cyanotype print by Munem Wasif, from the
series Seeds Shall Set Us Free, 2017–19. © Munem Wasif /
Agence VU'.

For Shafique and Suli

CONTENTS

MAPS

Winter 2010 found me in Bangladesh thinking about my next research project. After two decades of work on Pakistan, I was finally ready to do something closer to home. On a visit with my father to his village home in Manikganj, just a bit northeast of Dhaka, the capital city, we reached Aricha, a ferry terminal that routinely transported cars, buses, people, and animals across and up and down the Padma River. I looked on in shock at what lay in front of me. Used to a roaring river in my youth with ferryboats tentatively traversing it, I now saw a huge sand dune between the two banks and extending into both reaches of the river. A thin crack of water served as the passageway for the ferryboats, which now looked large and lugubrious to my eyes. My father, a soil scientist and ever my teacher, explained the hydrology of the river system in Bangladesh that led to the rivers being both destroyers and creators of land. While he marveled at this machine of nature, he said that the soil was too unstructured to be useful as roads, thus necessitating the ferryboats. However, he waxed poetic at the fertile quality of the soil, pointing out the rice paddies growing along the river's edges, likely planted by nearby villagers who would later transplant them elsewhere.

Used as I was to hearing about how Bangladesh was in the eye of the climate storm and stood to lose 25 percent of its landmass within a few decades to rising ocean waters, I was struck by this physical evidence of a countertendency within the landscape, and by the ready absorption of what the river threw up into the rhythms of everyday life. This was how I came to find my next research topic. I decided to study how people made lives for themselves alongside capricious rivers, specifically the Brahmaputra-Jamuna River, and the ever-shifting land that the river provided and whose status as a curse or a boon was never certain.

I decided to try to understand how the physical volatility of the riverine landscape was absorbed into the sinews of the social. The colonial and postcolonial history and political economy of Bangladesh went a long way in helping me to see how ragtag communities of itinerant farmers and fishermen came to be in these locations and to be economically vulnerable in very particular ways. I also found much, from gestures and feelings to sudden organization into patterned behavior as a group to flashes of intuition and senses of invisible forces, that could not be explained through the usual analytic frameworks. Although sometimes attributed to the omniscient presence of the divine through the language of the theological, very often such excess was referenced as simply a lure, an invitation, a pulse, or a presence, sometimes within oneself as much as an external cue.

This book is about giving an adequate description of this existence, without claiming for it the status of settled sociality. It is also about learning to ascribe authority to nature as one of the forces at play within this mode of existence, without allowing this to mean only the physical landscape and the human and nonhuman animals living in it. And it is about acknowledging that we still have to contend with nature both as concept and as alive in the world, or rather as concept precisely because it is alive in the world.

After I finished my first book (which took ages), I thought the next one would surely be easier because I now knew what the process involved. Little did I know that I would be changing not just my field site but also my focus, requiring me to seek out a second education almost the length of my doctoral training in order to research and write this book. For my new interest in riverine society I have first and foremost to thank my father, Shafique Khan, who inspired me to look at the riverine landscape through his eyes, to see his youthful escapades and subsequent travels as tied to the movements of the rivers that crisscross Bangladesh. I have countless memories of us making river crossings by ferry, of him pointing out sites, trees, and plants familiar to him, while he placated us children with snacks from the vendors who thronged the ferry ghats. So perhaps it wasn't so unexpected that I would find myself drawn to the same landscapes when it came time to dream up a new project.

I thank the Wenner-Gren Foundation, the American Institute of Bangladesh Studies, and the American Philosophical Society for providing funding for the first stretch of my research dating back to 2010. This was followed by generous funding by the Andrew Mellon Foundation New Directions Fellowship and the Johns Hopkins University Catalyst Fellowship, both of which afforded me the time and opportunity to study and learn to appreciate the rivers through the eyes of hydrologists, climate scientists, and German Romantic philosophers, and to return for repeat stints of fieldwork. I thank Peter Wilcock, Ciaran Harman, and Erica Schoenberger from the JHU Department of Geography and Environmental Engineering, Ben Zaitchik and Darryn Waugh from the Department of Earth and Planetary Sciences, and Eckart Förster of the Department of Philosophy for generously educating a colleague. While the earth and all its forces came into focus as I took STEM classes, I acquired an entirely new appreciation for the human as a modest mediator of such a vision of earth. This happened by way of an education in German idealism I received from Eckart over almost three years of coursework shoehorned into semesters in which I acted as both teacher and student. I owe a huge intellectual debt to him. I thank Jane Guyer, then the chair of the Department of Anthropology, for supporting me in this educational venture and my colleagues in the department for accommodating my leave.

Tahera Yasmin—Tulie apa to those of us who grew up with her as an affectionate, teasing, sisterly presence—introduced me to Habibullah Bahar, director of the nongovernmental organization Manob Mukti Sangstha (MMS), who warmly incorporated me into his life and that of his lovely family and facilitated my introduction to the Sirajganj chars, as the sandbars in the river were called, which came to be my home. I cannot thank Habib bhai and those who make up MMS enough for all the hospitality and care they have showered on me. Habib bhai's musings reverberate throughout the chapters. I appreciated very much MMS's sharing of Shohidul Islam; Shohidul came on as my research assistant but soon became friend and interlocutor. He and Mosarouf Hosain provided much-needed help in conducting surveys. Eventually, Shohidul bhai (brother), Kohinoor apa (sister), and Salaam bhai made up my small research team and family, with whom I spent countless wonderful hours in Dokhin Teguri walking about, shopping, boating, fishing, chasing chickens, cooking, and doing whatever else needed to be done to make daily sustenance possible.

I am not sure how to begin thanking the people in the chars who went from first treating me as very precious as a potential donor to their many life projects to giving up and impatiently folding me into their bosoms instead. I have rarely enjoyed such exquisite company and conversation as I had with chauras, as those who live on chars are sometimes called, and what I have learned from them could fill many books. I won't name them individually here, since this would require many pages, but instead point out that they are named in the chapters ahead as they asked that I use their actual names instead of pseudonyms. Their names will keep bursting forth from me for a long time to come.

The writing of this book has been a slow, iterative process through which I have had to learn to control my urge to speak to too many audiences at once. My editor, Ken Wissoker, and two anonymous reviewers have been patient and sure guides in making this book more honed and readable while helping me to clarify my stakes to better uphold my ambitions. I am very grateful for their guidance and their trust in my work. Lisa Lawley was a terrific editorial manager, shepherding the manuscript through the entirety of the process.

My constant companions through this writing process have been Andrew Brandel, Swayam Bagaria, Bhrigu Singh, and Sharika Thiranagama, the four most knowledgeable and generous friends one could hope for. Andrew was the first to introduce me to the strain of *Naturphilosophie* within German idealism, which I refer to in the book through the figure of Schelling, and he continues to impress on me the importance of literature for anthropology. Swayam, Bhrigu, and Sharika have been both perceptive and incisive in their comments on countless versions of my writing. My long-standing engagement and conversation with Veena Das, through her work and in the courses we have taught together on how Hindu and Islamic thought put pressure on anthropological concepts, are evident in many parts of this book. I am deeply grateful for her friendship and that of Ranen Das.

I thank, too, my colleagues Tom Özden-Schilling, Canay Özden-Schilling, and Michael Degani, who generously read and provided comments on an early version of the manuscript; Rochelle Tobias, my terrifically brave colleague and friend in German languages and literature, who jumped into the ethnography while checking my exposition of Schelling; and my former student and now colleague Vaibhav Saria, who generously read and edited the final version of the manuscript and whose encouragement has often lifted my spirits. Former and current graduate students who have read and provided helpful feedback on parts of the book include Nat Adams, Burge Abiral, Ghazal Asif, Sruti Chaganti, Amrita Ibrahim, Kunal Joshi, Sanaullah Khan, Perry Maddox, Sid Maunaguru, Basab Mullik, Sumin Myung, Maya Ratnam, Aditi Saraf, Megha Sehdev, Chitra Venkataramani, and Anna Wherry. I learn and grow so much through my engagement with students that it is hard to think of this book outside of our ongoing conversations. Friends and interlocutors at Hopkins include Jane Bennett, Bill Connolly, Debbie Poole (now retired), and Todd Shepard, for whose intellectual and political vibrancy I am very grateful.

Many friends and colleagues have provided ample opportunities over the past ten years to present my work at various phases or have given generously of their time in helpful conversations and intellectual nourishment. They include Vincanne Adams, Kamran Asdar Ali, Jonathan Shapiro Anjaria, Ulka Anjaria, Andrea Ballestero, Anna Bigelow, Dominic Boyer, Nils Bubandt, Iftikhar Dadi, Naisargi Dave, Maura Finkelstein, Karen Gagne, David Gilmartin, Ann Gold, Suzanne Guerlac, Akhil Gupta, Charlie Hallisey, Thomas Blom Hansen, Stefan Helmreich, Nathan Hensley, Cymene Howe, Robin Jeffry, Arthur Kleinman, Franz Krause, Sarah Lamb, Stephanie LeMenager, Dana Luciano, Purnima Mankekar, Setrag Manoukian, Andrew Matthews, Stuart McLean, Diane Nelson, Ben Orlove, Katrin Pahl, Priti Ramamurthy, Mattias Borg Rasmussen, Mubbashir Rizvi, Liz Roberts, Lotte Segal, Martha Ann Selby, Harris Solomon, Anna Tsing, Bharat Venkat, Ara Wilson, and Vazira Zamindar. I befriended Kate Brown and Danilyn Rutherford at a workshop and Anant Maringanti at another; they have been the best finds. Scholars of Bangladesh—all friends and fellow travelers who are my spur to ensuring that my contributions to the scholarship on the country are as rigorous and critical as they are sympathetic—are Tariq Ali, Kazi Khaleed Ashraf, Jason Cons, Lotte Hoek, Iftekhar Iqbal, David Ludden, Naeem Mohaiemen, Kasia Paprocki, Ali Riaz, Elora Shehabuddin, Dina Siddiqi, Tony Stewart, and Willem van Schendel. I have enjoyed being in conversation with Seuty Saber and Samia Huq over emergent issues of anthropological concern in Bangladesh, while Firdous Azim, Fazlul Haque Majumdar (aka Ripon bhai), and Saymon Zakaria have opened up aspects of the Bengali literary, poetic, and folk for me through their preoccupations. Chapter 5 was previously published as "Living Paradox in Riverine Bangladesh: Whiteheadian Perspectives on Ganga Devi and Khwaja Khijir" in *Anthropologica*. I thank the journal for allowing me to use the ethnographic material.

Looking homeward, I have loved being able to adda with my childhood friends Aleeze Moss, Dina Hossain, Lopita Huq, Lamiya Morshed, Seema Karim, and my fabulous sisters, Shaila and Sabina, who are nonplussed by what I do but are nonetheless my biggest cheerleaders. The pleasure of being able to go between my home in the chars and the joyful home of my parents, Munawar and Shafique, and my brother, Shahed, and sister-in-law, Rumana, in Dhaka, after such a long stretch of living outside of Bangladesh has made this fieldwork more special than anything I have done or will do. Finally, all credit for this work lies with my resilient children, Sophie and Suli, who bore my absence during long stretches of fieldwork with heroic stoicism, and my wonderful husband, Bob, without whose support I doubt I would know what it is to be a scholar in my own right. If there is anything that these past few years of political turmoil have taught me, it is that we cannot take any advances on rights for granted, and that it takes many people to make one person's journey a success. As Jane Guyer would say, I am rich in people.

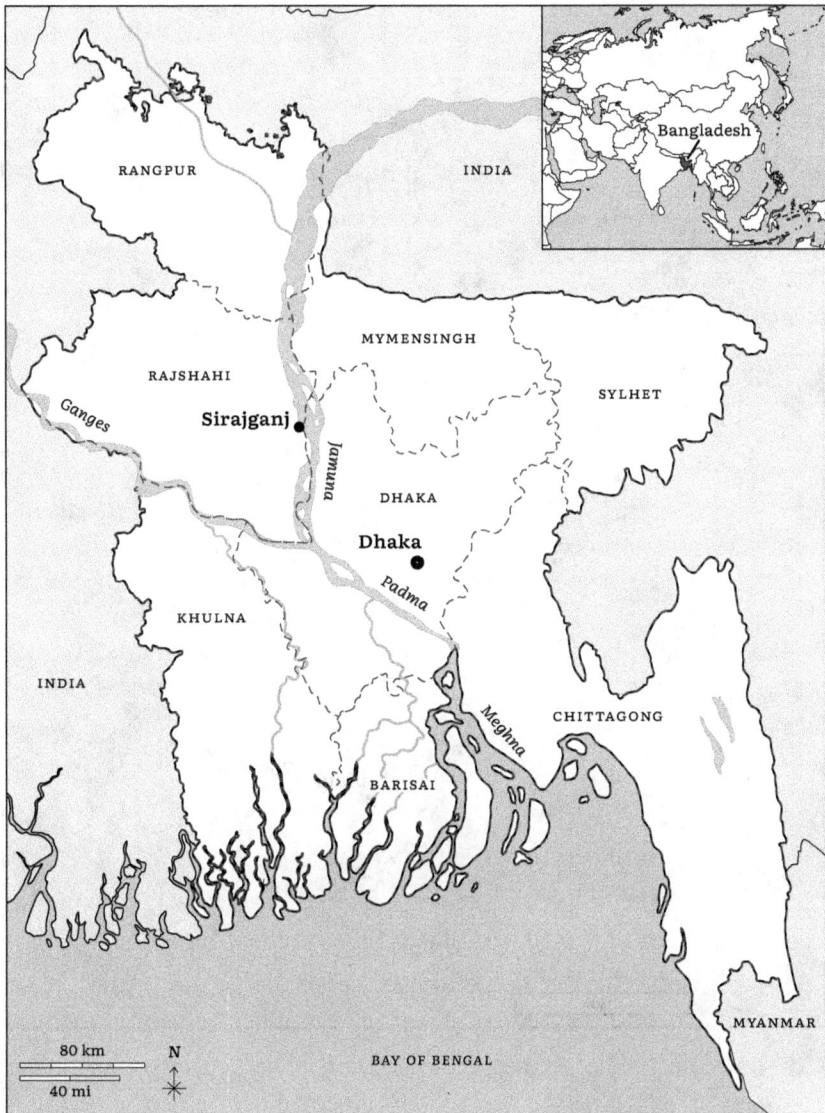

Map 1. Divisional map of Bangladesh.
Source: author created.

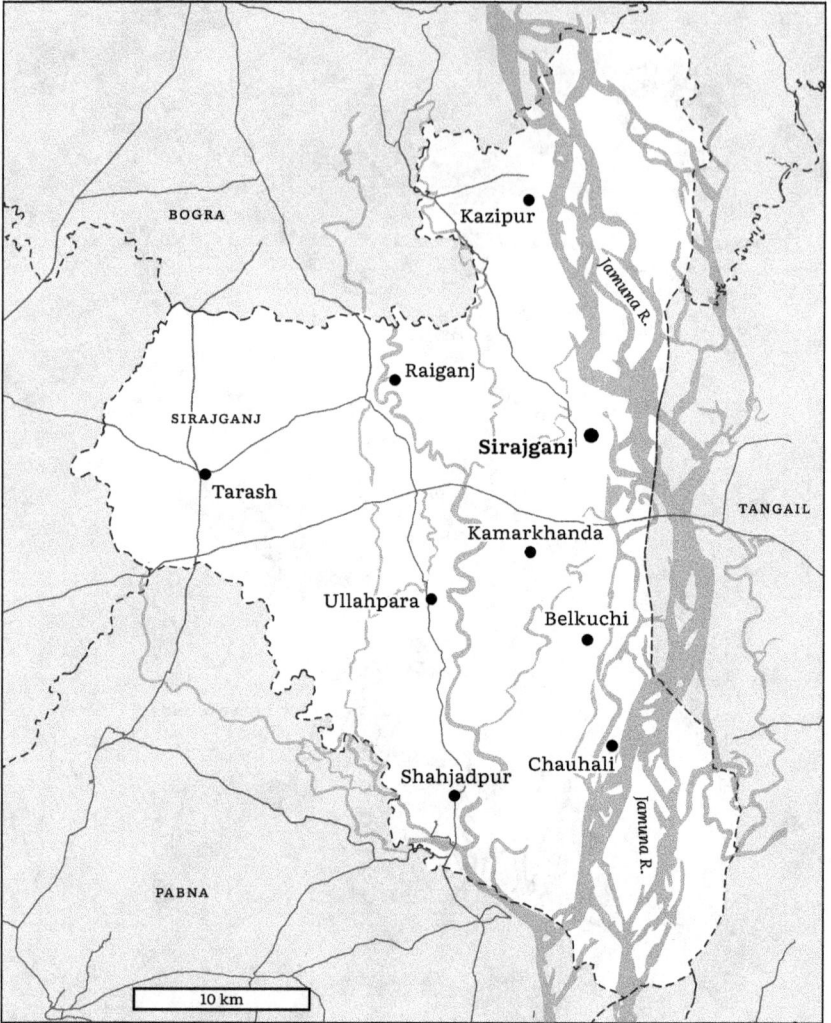

Map 2. District map of Sirajganj.
Source: author created.

Map 3. Satellite image of Chauhali.
Source: Google Earth.

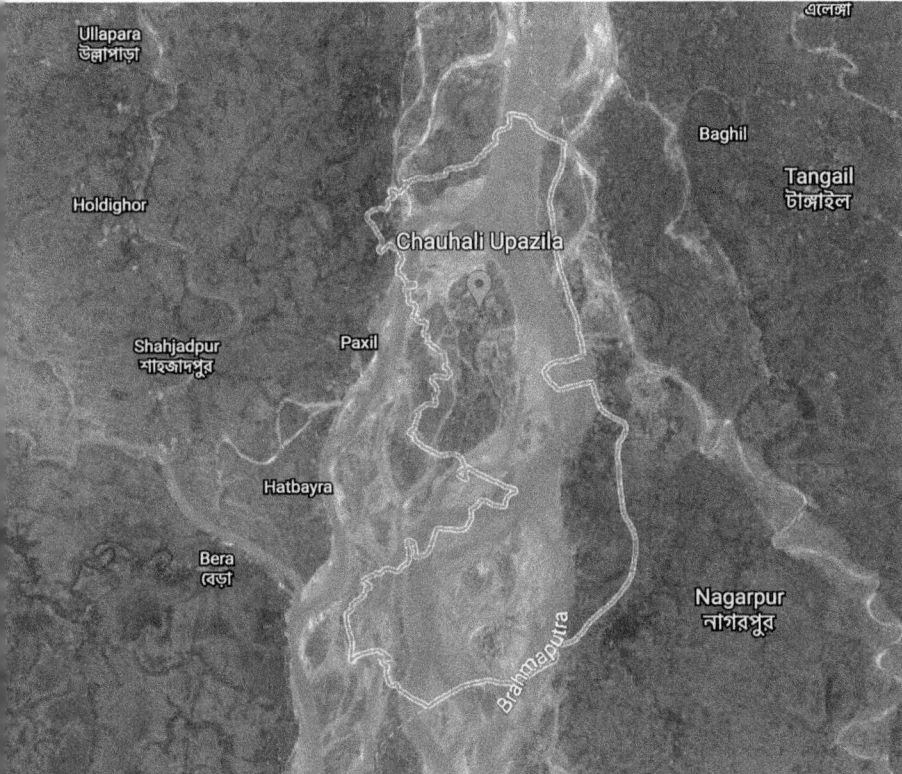

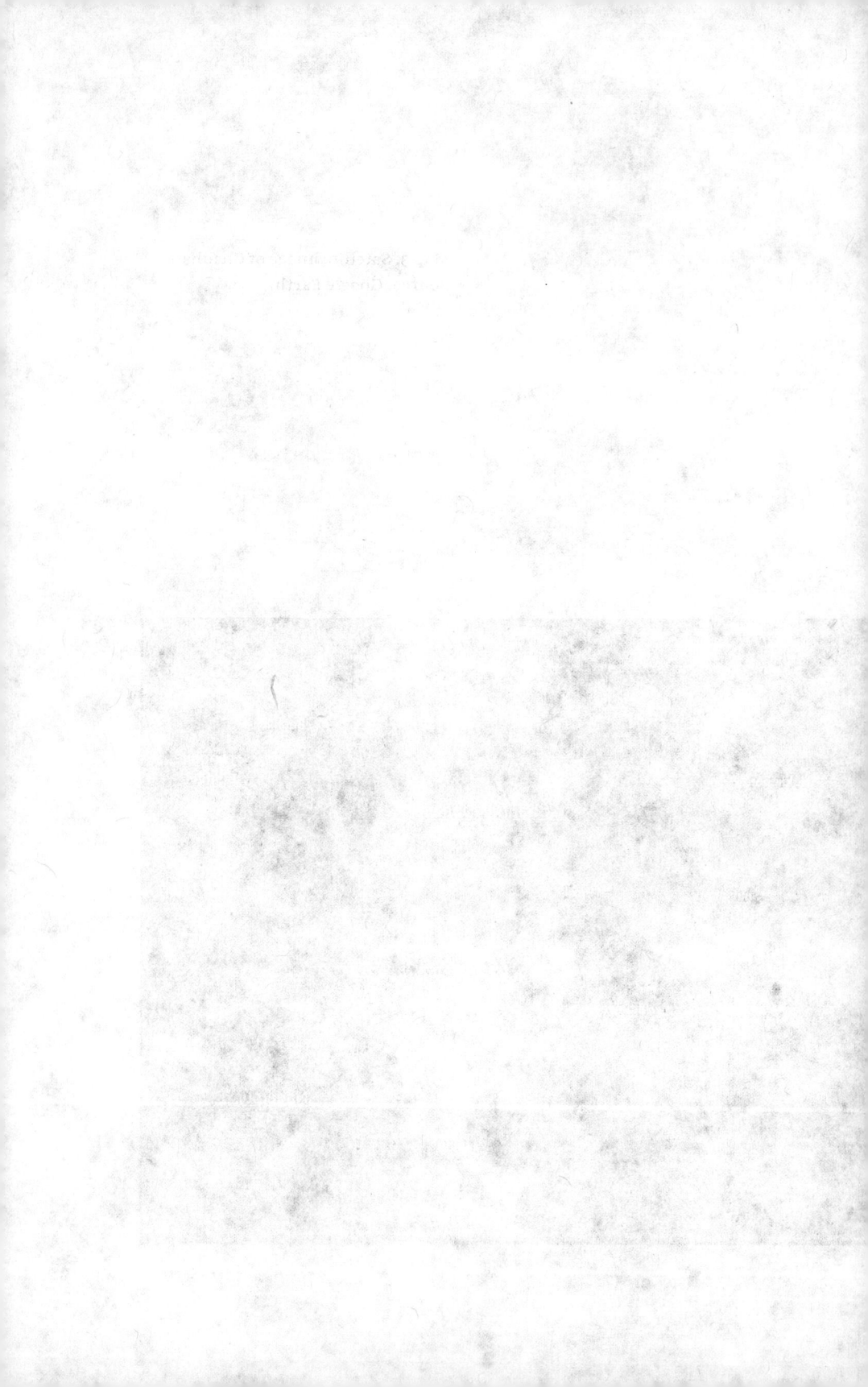

INTRODUCTION.   *River Life*

                                        and Death

THE JAMUNA RIVER ORIGINATES IN TIBET as the Tsangpo; breaks through
the Himalayan mountain range in great gorges into Arunchal Pradesh,
India, where it is known as the Diang; flows southwest through Assam as
the Brahmaputra and enters Bangladesh through the north as the Jamuna,
whereupon it first merges with the Teesta River; and then merges with the
Padma River and finally with the Meghna River before emptying into the
Bay of Bengal and the Indian Ocean. Again and again I heard that this was
a complicated river system and not only because it flowed across so many
contentious nation-states and served as the drainage system for an area of
583,000 square kilometers. Its complexity further derived from the fact that
its sediment load and the low slope of the Bengal Delta made it a braided
river (Coleman 1960; Best et al. 2007; N. M. Islam 2010). In other words, it
was not a river with a definite channel and a forceful flow. It was made up
of many subchannels, or braids, that scuttled indecisively back and forth
across the landscape even as the water in them was pulled inexorably by the
tides of the Indian Ocean. While the braids added considerable variability
across the Jamuna's length and breadth, snowmelt and monsoon rains also
made the river widen and contract seasonally between 3 and 10 kilometers,
in some places reaching 20 kilometers, and occasionally changed its nature

to have a more decisive flow, overspill as floods, and channel migration through literally leaping over to another location in a process called avulsion (Sarker et al. 2014). The most interesting element of the system to those who study such phenomena was that the river transported a heavy load of sediment that it deposited wherever the water lost its force of will to carry its load, which was usually downriver, creating land along the banks and in the center of the river. Just as often, if not more so, the river, through its increased volume of water, acceleration, or overspill, eroded the very land that it produced. For those who lived downriver, the river, thus, became the giver and taker of land (Sarker, Huque, and Alam 2003; de Wilde 2011).

In 2011, I entered life on the delta at the location in the Jamuna River between the districts of Sirajganj and Tangail, known for its high rates of accretion and erosion (Bangladesh Water Development Board 2010), to study the lives of those who lived on the chars. Those living on the mainland used the term *chaura* as a derisive label for them, but those living on chars took up the term with amused alacrity, referring to themselves as *chaura manush*, or people of the chars. Although of Bengali ethnicity, chauras came from diverse lineages and occupied many different communities within the villages on the chars. They were neither entirely homogeneous nor quite as motley as mainlanders made them out to be, as will become evident over the course of this book. I decided to use the term *chauras* interchangeably with *char dwellers* to indicate this population whose backgrounds and behavior were within a given range of variation, as well as to indicate an existence that carried the sting of mainstream judgment.[1]

During my very first visit I was quickly alerted to the fact that chars held an illusory quality, even for chauras. From the boat approaching island chars, it took many rotations of the head and shifting of the body to tell apart the glimmering water from the onset of land with its reflective sandy surface, which lay flush with the water. Within a few months of living on a char, I became attuned to the fact that the village where I was staying was perhaps in its fifth incarnation. As I walked across the sandy stretches that separated one village from another to reach the bank from which to take the boat to the mainland, I was told by my walking companion that movement was easier when one could take a boat directly from an earlier version of our village to the mainland, that the newest version of the village was very inconveniently located vis-à-vis the mainland. In other words, the rainy season alone did not determine whether one walked or boated to the mainland; the changing location and surrounding topography of the village did as well. That many past emanations of the village were intermingled with the present was repeatedly made clear to me through gestures toward physical locations that had held excellent fruit trees several versions of the village ago, or toward lowland that had once been ideal for cultivation but returned as upraised land better suited for settlement. Many pasts pressed upon the present. And every action undertaken in the present had many future possibilities built into it. One of my closest interlocutors once described how she planned her home garden so that, should it rain, the seeds she had planted near her house would grow into trees, but should it flood, the floodwaters would transport her seeds to the pit that she had dug close to the house for such an event, and should the land break, she would scoop out the young stalks to take with her to wherever she went next.

Chauras lived in a conditional mode, with many "if this, then that" scenarios crowding their daily life and future horizons. The issue of the interrelation between chauras and variable temporal horizons takes on urgency in the era of global climate change, when overwhelming evidence suggests that human activity is forcing changes to the many processes that produce the global climate. This has consequences not just for daily weather and seasonal variation but also for ocean waters that are considered to be acidifying and rising, which, along with glacier melt, could have potentially catastrophic results for a low-lying deltaic country such as Bangladesh. While rising ocean waters threatened to inundate Bangladesh's coastal lands, its river pathways threatened to carry salt water upcountry (Mirza, Warrick, and Ericksen 2003; Al Faruque and Khan 2013; Gain et al. 2013; Brammer 2014).

At the same time, the precise impact of climate change is hard to pin down in a physically dynamic system such as the Jamuna River (T. Islam and Neelim 2010). For one thing, the river in its current configuration is the reverberation of the events of past earthquakes in the lower reaches of the eastern Himalayas (Sarker and Thorne 2006). The last earthquake in 1950 in Assam, an Indian state due north of Bangladesh, caused entire mountains to shake and collapse, producing a tremendous pileup of sediment blocking all waterways. In response, the Jamuna widened, deepened, and proliferated its branches to carry the sediment downstream. The river was only doing what it was created to do, for it was another such earthquake in the late eighteenth century that had abruptly shifted the Jamuna's course southward from its earlier circumscribed and leisurely route northwest. And should there be no more earthquakes, then the river could simply run its course once the sediment buildup from the 1950 seismic event has been transported. The river is a trace of a geological event coming to an end.

In effect, the chauras were caught in the conjunction of two temporal sequences, one from the past with the river as the reverberation of earthquakes and the second from the future as the ocean with its rising waters—warm, salinic, and acidic—is slated to enter the river system.[2] The likelihood that the sediment transport could decisively cease, thereby stopping the engine of the river, or that ocean waters could reach the upper reaches of the river, puts into question the very viability of chars. This situation allows us to speculate that the chauras were living a form of life that was likely coming to an end.

Chauras in Context

The district of Sirajganj was no more than eighty miles outside of Dhaka, the capital city of Bangladesh. Yet to reach my field site took upward of five hours, as I had to take a train to Sirajganj Town, then catch a ride to the riverbank closest to the char where I stayed, then a boat across to the bank closest to the mainland, rounded out by a motorbike ride over sandy dunes, shallow rivers, and frail upraised pathways between cultivated fields to arrive at the field office of the NGO Manob Mukti Sangstha (MMS) or the Organization for the Freedom of Humanity suboffice where I was provided a room of my own. The bank on both sides of the char kept changing, and so my point of departure and arrival kept changing until finally I realized it was best for everyone when there was water everywhere because then we could travel by boat. But this was not to be as Sirajganj was constantly undergoing repair to secure the buttresses holding back the water.

Sirajganj first emerged out of the waters as land accreted to zamindari (landed gentry) properties in 1884 in colonial India and only became a district in 1984 in independent Bangladesh. It was administrative will that had deemed Sirajganj would be a district as it had a frail spatial existence with no fewer than five main rivers and numerous branches across nine *upazilas*, or subdistricts, all of which were once char lands or were in the process of becoming chars. Once part of the isolated and impoverished northern half of Bangladesh, Sirajganj became better integrated into the country after the construction of the Jamuna Multipurpose Bridge between 1994 and 1998. Due to budgetary constraints, the bridge was not made as wide as it needed to be, necessitating ongoing infrastructural repair and river training to keep the water flowing under the bridge. One such effort has been to shore up Shirajgank *shodor*, or the town of Sirajganj, which had once been a flourishing site of the jute industry in colonial India but was abandoned once the river around it became too silted to allow the passage of boats. But since that time the river had changed course, leading the town to be ravaged repeatedly by it. Securing the bridge meant securing the town, which meant building a hard point, a short stone dike used to stabilize river banks, along the river's eastern bank between 1995 and 1999 to stay river erosion. The hard point had been compromised numerous times starting in 2009.

Between 2011 and 2017, I lived and worked intermittently in one of the larger chars in Chauhali subdivision due south of the Sirajganj Town hard point and the Jamuna Bridge. Although these were the very chars faulted for causing trouble for both the town and the bridge, it was nearly impossible to imagine how they could be scoured out of existence because they not only were massive in scale but also held large populations. The island on which I worked was wedged in a branch of the Jamuna River between Sirajganj and Tangail, its neighboring district. It was ten square miles in area and housed no fewer than ten villages with anywhere between 50 and 250 households each, with each household of an average size of five to seven people. The government just let the chars be with no effort to remove them but with also no effort to strengthen them to withstand the river or provide basic services to their inhabitants. When char dwellers said they lived in "the remote" (using the English word), they did not mean at a great physical distance from the center of government, because Sirajganj Town was close to Dhaka. They referred to a sense of remoteness produced of neglect by the central and district government, apparent in the lack of electricity, roads, and schools, the three major markers of development evident elsewhere in the country.

Living in the shadow of major infrastructural projects, which chauras neither condemned as the Jamuna Bridge had brought economic prosperity to the previously isolated Sirajganj nor condoned as these projects, however desultorily, sought the demise of chaura lives and land, they made the best of what they had with little expectation of government help. Rather, family, village, and political ties had to be maintained and thickened to the extent possible to tug on as situations arose. And situations arose quite often, as chars alternately faced deluges and droughts, with large chunks of land eroding into the river.

Early in my fieldwork, I chose three villages on the island, on different topography and at different stages in the process of erosion and accretion, to get closer to the grain of the chaura everyday. The village of Dokhin Teguri, or South Teguri, had been around the longest, having emerged after the floods of 1988, and was still going strong in 2011. This long duration was unusual for a char village, secured no doubt by the bridge immediately upriver. The village had birthed a generation of children who had never experienced river erosion and who thought of themselves as living on *qayem*, or established land, while living in the middle of an erosive river. I was interested in how this sense of durability and permanence was maintained in the face of its history and likely future. This village served as my home in the chars.

Rihayi Kawliya, or Mercy on Kawliya, the second village I studied, was almost gone, eroded by the river. Only a mosque and a few households remained, but those who lived near it on borrowed and rented land from adjoining villages spoke as if Rihayi Kawliya were still there around them, while the majority of their co-villagers who now lived on the other side of the river in Khas Dholayi in the district of Tangail persisted in saying that they too lived in Rihayi Kawliya. While the actual village of Rihayi Kawliya might have gone, it existed virtually on both sides of the river, and it was this village in absentia that I studied to understand the work that went into keeping it present and making its presence count.

The third of my village sites was a new wing of the village of Boro Gorjan, or the Big Roar, composed of those who had taken the boat over posthaste from a nearby subdistrict as three villages in that area collapsed into the river. In the initial stages of their settlement the area was called Hotath Para, or Sudden Neighborhood, to indicate the suddenness with which it had gone up. The inhabitants of Hotath Para, later renamed Kuwait Para, lived cheek by jowl, facing inward into their households with their backs resolutely to the riverbank. With their villages gone, they had to forge a precarious sociality among strangers, perhaps similar to the country of Kuwait that is home to many migrant laborers, while banking on a future when their lands would return.

At one time, in the colonial and more recent postcolonial past, the char dwellers were considered an untrustworthy, rootless people (Baqee 1998). It was as if their character was a direct reflection of their lands, which broke and re-formed with such regularity and intensity as to leave the soil churning in the waters to be watched by the elderly who congregated along the riverbanks during the evenings. This continual movement of land and people kept the British colonial land laws in the nineteenth century in a state of constant amendment until such point that the settlement officers declared, "We think it impossible to lay down any fixed laws for a shifting sand," and the colonial government decided to survey and settle only those char areas not of "a fluctuating nature" (Hill 1997, 47). In the fluctuating areas, no records of rights were to be published, and no settlement of revenues or rents was to be made. This left chars in an indeterminate legal state that continued through the partition of 1947, the period of East Pakistan between 1947 and 1971, and into contemporary Bangladesh, resulting in the endemic violence and fraught sociality that marked char life (I. Iqbal 2010).

It was also the movement of chars and their indeterminate legality that made them home to one of the longest-standing peasant rebellions against the colonial government in the nineteenth century, the Fairazi Movement (I. Iqbal 2010). The rebels relied on the relative lack of presence of the colonial state in chars to sustain and organize themselves. The waterways served as their means of movement and communication. The Fairazi imagination of an egalitarian community with equal rights to the land derived not only from peasant interpretations of the tenets of Islam but also from the nature of chars to form and re-form, producing the effect of a tabula rasa, the chance to start afresh. Chars eroding decisively and accreting equally dramatically helped sustain the imagination of divine will independent of human will but also sympathetic to human rebellion against oppression, insofar as chars lent themselves to organizing outside of the surveillance of the British colonial state. It was not uncommon for Fairazi revolutionary pamphlets to point to rains turning into deluges in these parts to the detriment of British authorities and zamindars as divinely ordained punishment.

In postcolonial Bangladesh, with land settlement stalled at present in acknowledgment of the difficulties of fixing land in these parts and of controlling illegal activity monitored by the state, the national newspapers and development agencies had moved away from describing the char dwellers as inherently shiftless to portraying them as among the 24 percent of hardcore poor in Bangladesh (see, for instance, Brocklesby and Hobley 2003). They came into particular focus during natural disasters, portrayed either as fatalist for ascribing their condition and misfortunes to Allah's will or as resilient against the elements (Indra 2000; Schmuck-Widmann 2000, 2001). Yet attention to chaura speech suggested that they were not quite either, that there was a specific mode of engagement that neither railed against nor was passive in the face of natural disasters. For instance, although they accepted the inevitability of natural disasters as an erratic element in the world, in the aftermath of these events, the char dwellers sought to reinsert themselves into their physical surroundings—for instance, by pulling their houses to higher ground, searching out seeds to plant within a given time, or dragging weaving machines to a dry location to meet a deadline for a shipment of woven cloth. In other words, they let the erratic quality of disasters buffet them and then worked to step into rhythm with the diurnal and the seasonal. And if nature was to be found in the chars in the many physical stirrings of matter, it was also to be found in the conjoining of such physical dynamism with receptivity and attentiveness toward the world to eke out repetition and regularity from it.

While it took work for the chaura everyday to acquire regularity and repetition, this tempo suggested neither a mastery over the physical surroundings nor something temporarily seized from the environs but rather a chaura mode of shifting grounds in relation to moving lands. In other words, the various social institutions and mores that constituted chaura lives evinced not only a flexibility to accommodate a wide range of circumstances and possibilities but also flexibility about the agreements that constituted the social. For instance, while strict apportioning of land by ownership held when villages were in their right places, when lands eroded or were newly up, everybody farmed collectively whatever land remained or any new land, sharing the proceeds according to how much labor was expended by each person. Or, for instance, when lands were secure, villages evinced a clear political hierarchy similar in nature to the villages in rural Bangladesh. But when the lands were no more, the households in their temporary locations operated entirely as individual sovereign units giving little to no credence to prior village leaders, as now they held the same status. While the chauras maintained a strict standard for women's privacy in their households, when their village was no more, it was as though keeping purdah was inconsequential. Women and men went about their business of securing livelihoods and making do while living by the roadside until their lands reemerged or they found new land on which to put up their households. It wasn't merely the case that circumstances constrained or even prevented char dwellers from assuming the conventions that usually shored them up as a social group; rather, different understandings of ownership, labor, hierarchy, and gendered norms held for different circumstances.[3]

The Upspring of Nature

These multiple temporal horizons, shifting grounds, and disparate agreements that made up chaura lives led me to study how moving lands did not simply detract from but in fact enabled a chaura mode of existence. Yet I was unhappy by how often I was read as saying that the landscape forged this form of life, as if there were a simple relationship between the landscape and chaura lives, with the landscape providing the external determinants for chaura experience. This was the kind of naive materialism that could lead to geographic determinism and the racialization of place that the best of anthropology was leery of and had led it to maintain a sharp distinction between nature and culture. This formulation also seemed to miss the fact that while the moving lands indeed forced the chauras to live under very difficult circumstances, the chauras were also drawn to this moving quality of land with its inherent promises and threats. The physical dynamism allowed the flourishing of different imaginations, norms, and even ideals among the char dwellers.

I understood the join between landscape and people, nature and culture, a bit differently, taking it to be more intimate. It is the claim of this book that chaura lives are configured by nature, that nature makes persons and cultures in this place, as it likely does in every place, but one sees the productivity of nature more clearly here. While we have long acknowledged that nature provides the material conditions for existence and informs corporeal life and embodiment, it has only ever been thought of as an external spur. My argument is that nature is more internal than external, more subject than object, and serves as the ground and possibility for human subjectivity, thought, and culture.

There are weak and strong versions of this view of nature. Just as we have come to think of the nation-state or the free market as abstract entities with distinct histories, existence as ideas, and material effects, the weaker version maintains the same for nature. R. G. Collingwood (1960) and Pierre Hadot (2006) have tracked the idea of nature, while Philippe Descola (2013a, 2013b) has shown how this version of nature has acquired the status of ontology. Just as the nation-state may produce feelings of belonging and capitalism may constitute us as desiring beings, so too do ideas of nature produce us as particular kinds of persons, to be explored over the course of this book.

The stronger version of this view of nature, which I pursue in the book, claims that although it is available for human observation, seen as intrinsic and necessary for human existence, and theologized or anthropomorphized in turn by humans, nature is unconditioned. As a force, tendency, or orientation, nature actualizes itself within the human mind, body, and modes of existence but remains alien and enigmatic in the sense of being exterior to human cognition and indifferent to human will and desire.

We may get a contrary perspective from Bangladesh studies, which urges us to focus on the specificities of state-society relations in Bangladesh in shaping chaura realities. Or anthropologists may protest, having long battled other disciplines for making simplistic and even dangerous statements about how culture is an adaptive response to the needs of physical survival and biological self-perpetuation. I take up each of these perspectives in turn to draw out their merits for thickening our descriptions of chars, but also to show how none fully suffices at acknowledging nature within our accounts of the diverse modes of existence that I find in chaura lives and articulations.

While here I don't rehearse the entire field of Bangladesh studies (see N. Khan 2015), two trends are particularly relevant with respect to framing the lives of char dwellers. The first has to do with the historical formation of Bangladesh as a nation-state and the issues that ensue from that. The second entangles char dwellers through the focus on development as the dominant paradigm by which the country crafts its future.

Unlike India and Pakistan, for which the partition of 1947 played an important role in shaping their respective histories and historiography, it was 1971, or the year Bangladesh gained its independence from Pakistan at the conclusion of a bloody war between the Pakistani army and Bengali liberation fighters, that constituted the founding event for Bangladesh (Riaz 2016). This event has been examined from many angles, from its roots in West Pakistan misgovernance of East Pakistan; to the successive suppression and commemoration of 1971 by Bangladesh's ever-changing governments, which largely oscillate between the two national parties, the Awami League (AL) and the Bangladesh Nationalist Party (BNP) (Mookherjee 2007, 2015); to its paradoxical inheritance by the young, ranging from those who are committed to the secular principles that informed the country's early constitution to those who are more attracted to an Islamic ethos (N. S. Chowdhury 2019).

The chars and those who dwell on them were very much part of this national story due to the fact that these chars were part of the mainland during the 1970s and bore the brunt of the offensive on the countryside by the Pakistan army. Chauras recalled their suffering at the hands of the armed forces and the effects of West Pakistan's scorched-earth policy, with a number of them describing their days as part of the guerrilla force fighting for liberation. The year 1971 and its aftermath, specifically the pains of reconstruction, the unfulfilled promise of land reform, the 1974 famine, the failure to try those who collaborated with Pakistan's army, the coups, and the continual tussle between the two national parties, reverberated through their lives (also see van Schendel 2009; Ali 2010).

Within this national story the chauras are rendered generic Bengalis who suffered Pakistani oppression and who now threw in their lot with the AL political party, viewed as committed to seeking justice that had been deferred after independence. While those living in these parts overwhelmingly supported AL, this story could not possibly encompass chaura memories or realities (N. Khan 2021b). After all, char dwellers lived not with one stable past but with many pasts of varying temporal depths pressing into their present. Furthermore, entities and events of greater age than the fifty-year-old state of Bangladesh moved through this mode of existence. There was little in the history of Bangladesh to accommodate the conditionality that informed chaura lives or the forces to which they gave expression. A disacknowledgment of these dimensions of chaura lives was also a suppression of the fact that the entirety of Bangladesh was geographically a char, an accretion on the landmass that constitutes the Indian subcontinent (H. E. Rashid 1991).

Since its independence, Bangladesh has been the poster child for international development. As Naomi Hossain (2017) has written, Bangladesh is known internationally as an "aid lab," for having served as the site of experimentation of diverse market-friendly policies, from the early dismantling of the food rationing and distribution system, to the green revolution, to microcredit and social enterprises that were underwritten by international funding agencies and executed by diverse NGOs and governmental bodies (see also Lewis 2011). While at times the char mode of existence had been slated for complete excision through allowing chars to be eroded, or even purposefully destroyed to make way for megaprojects such as the Jamuna Bridge (Penz, Drydyk, and Bose 2011), char dwellers had largely experienced only neglect by the government. It was only at the beginning of the twenty-first century that the chars where I worked became the focus of international development projects (Brocklesby and Hobley 2003), but those had since ceased as funds had dried up and the river swept away many of the projects along with char lands. The few development studies on the chars and char dwellers focused largely on the difficulties of life lived with river erosion and the resilience and vulnerabilities of those who encountered it, oscillating between living near the river's edge or else on roadsides (Haque 1988; Zaman 1996; Indra 2000). Other studies of char communities from an explicitly political economy perspective point to the incompleteness of land survey and reform, disallowing chauras from having the autonomy and political might that comes from holding property rights, even if these rights extended to land that existed only sporadically (Barkat, Zaman, and Raihan 2001).

As in the case of nationalist history and historiography, development studies attempts to capture and represent existing dimensions of chaura lives and realities. Those who dwelled on chars were as desirous of the opportunities that development projects promised as everyone else. However, the development paradigm too failed to provide a full understanding of what such lives entailed. There was a fundamental incapacity to appreciate any attraction or plenitude at work within chaura lives bred by the common sense that development was the only good. Although chauras very much lived in the political and socioeconomic space of Bangladesh as a nation-state, their lives and expressions, however disparate and inchoate, exceeded the space-time of Bangladesh and required further frames of reference.

While Bangladesh studies might not put up too much struggle in considering chars through a different set of lenses, after all, Bengali poetry and music speak eloquently of entities and forces beyond those of the nation-state alone, sociocultural anthropology's suspicion of any explanation for culture and society that takes its orientation from anything called nature is deepseated. While it may be foolhardy for me to attempt a schematic overview of anthropology's repulsion of any argumentation by means of nature, I still venture to do so because it helps demonstrate how necessary such vigilance has been to keep at bay any naive materialism with its dangerous entailments of the kind I mentioned earlier. At the same time, this has meant sustaining a nature-culture divide for longer than has been productive for anthropology. A few early examples should suffice to illustrate my point. In *Seasonal Variations of the Eskimo*, Marcel Mauss ([1950] 1979) undertook an effective critique of the prevailing notions of anthrogeography espoused by the German geographer Friedrich Ratzel, for whom states and societies were the organic outgrowths of their natural environments. While acknowledging that environmental factors were important, Mauss showed that there was no one-to-one correspondence between environment and society. Society's complexity far exceeded its environmental setting. Franz Boas waged a long intellectual battle against the eugenics movement in the United States using an early version of the argument that race was a social construct that had been naturalized, an argument carried forward by scholars such as Jonathan Marks, among others (Boas [1940] 1982; Marks 2017). Marshall Sahlins's scathing critique of the recourse to biology and ecology to study human society can be found in his Use and Abuse of Biology: An Anthropological Critique of Sociobiology (1976), in which he pushed back on E. O. Wilson's efforts to read out social organization from ant colonies to human societies.

Even environmental anthropology, which emerged and diverged from the mainstream of Boasian anthropology, was careful to not resort to crass materialist or biological explanations in its studies of societies. Both Leslie White ([1959] 2016) and Julian Steward ([1955] 1972), for instance, espoused an understanding of evolution that diverged sharply from the Spencerian and Darwinian understanding of it by bringing in the idea that in addition to physical environments, cultures in their growing complexity produced environments of their own that looped back into the evolutionary process. At the same time, their focus on evolution, over the Boasian preference for historical diffusion, as the paradigm by which to study the development of societies allowed for a greater focus on the relationship between environments and cultures. Roy Rappaport's Pigs for the Ancestors ([1984] 2000) was an ambitious attempt to bring together these two branches of anthropology through arguing that ritual had its own place and explanation immanent to a society, but also had utility in regulating the larger ecosystem. But Rappaport faced pushback on the grounds that he treated ritual, and culture by extension, as epiphenomenal to the maintenance of the ecosystem.

This quick schematic sketch of the state of debate on culture and nature within the early years of anthropological scholarship is to point out that the defense against nature was necessary, but also circumscribed nature almost entirely to the physical environment, race/biology, and ecology/ecosystem. This suspicion of nature has resulted in an exhausting focus on all things human to the exclusion of everything else. Thus, Donna Haraway's (2003) concept of natureculture was pioneering in bringing in the more-than-human aspect of human existence. The move away from standard narratives such as that of domestication to multispecies entanglements allowed for attention to the many nonhuman agents involved in the crafting of societies and cultures (Cassidy and Mullin 2007; Kirksey and Helmreich 2010; Tsing 2015). And although ontological perspectivism is much more informed by a commitment to thinking Indigenous thought as philosophy than to the more than human, it reverses the older dichotomy of one nature and many cultures to thinking about one culture and many natures, offering the insight that each species thinks of itself as human and its prey as animals (Castro 2014). Finally, vital materialism (Bennett 2010a, 2010b, 2020) and actor-network theory (Latour 2005) expand the scope for participation within the societies of humans and nonhuman animals for inanimate matter and objects. Elizabeth Povinelli's (2016) explication of geo-ontologies underlines the importance of paying attention to the inanimate in the various ways it is both included and excluded to map contemporary formations of power.

Such, then, are the voices that will likely protest from within anthropology that there is no call for me to revert to the concept of nature and to think of persons and cultures as products of it. The concept is overly burdened by a history of problematic associations and usage, and other generative frames of reference are available with which to bring into view the part played by the many creatures with which char dwellers make their lives, and the animateness and agency of the matter around them, such as that of the river and the lands on which they live. While there are critiques of these more inclusive approaches, such as that they may have evolved more out of a wishfulness to have societies be more participatory and democratic than the actual realities on the ground (see Morris 2017), in my reading they do not go far enough in making nature a felt reality both cognitively and existentially. Restricting themselves to nature as material, substantive, or embodied—that is, as environment, biology, systems, animals, matter, or objects—they leave out of consideration the aspects of nature as ideational and dynamic, as productive of subjects and consciousness as of species and rocks.

While Claude Lévi-Strauss was interested in thinking as a process immersed in and emerging out of the world, he made it independent of any association with either nature or culture, as hovering disembodied over both (see Lévi-Strauss 1962, 1992). Eduardo Kohn (2013) takes up Gregory Bateson's (1972) provocation to consider thought as being looped through the world through sign systems. However, his picture of the natural world is still one of an external object to which one has a frontal relationship, which is primarily that of communication. Henri Bergson's concept of "life," to which he turned in his *Creative Evolution* (1911) so as to be done with extant debates between idealism and materialism, captures the dynamic of nature as that which is simultaneously material and most ideational, most external and most internal, but restricts its scope only to the organic. Animism, or the attribution of soul to plants, objects, and physical phenomena, may have been a productive line of analysis for me to pursue were it not for the fact that there was such an abhorrence of any hint of pantheism within the predominantly Muslim milieu of the chars (Manzoor 2003). The chauras were as given to naturalism as any Enlightenment thinker.

Ultimately, it is only "nature" that can convey all these various aspects, not as a catchall term for everything, such as environment, biology, systems, animals, matter, objects, or thought, but as itself. Nature is not that which has to be overcome and mastered, or with which to be reconciled, but that which springs up within us and to which we offer our attention and receptivity, and activity and passivity as response.

Schelling and the Char Dwellers

Such a perspective on nature meant keeping humans in the picture, displaced from the center that they occupied in anthropocentric accounts of nature, but still a means by which nature expressed itself. Consequently, in my fieldwork I privileged the study of humans over that of the more than human. And given the thread between the chauras and the Enlightenment era through their mutual attraction to naturalism, it did not feel out of place to draw on the writings of Friedrich Wilhelm Joseph von Schelling, an Enlightenment figure, to help me navigate how nature made human consciousness the means for its own ends. Although I focus here on Schelling, I request the reader to imagine him to be a bit like a milieu himself, with much intellectual back-and-forth between diverse figures, specifically those who came to be called romantics and who, far from being irrational subjectivists, insisted on the rationality and impersonality of the tendencies and forces running through one (N. Khan 2021a; Nassar 2013).

Schelling is of the generation of German philosophers who came into prominence in the waning years of Immanuel Kant's life in the late eighteenth century. He was among those who appreciated the full scope of Kant's project of securing reason from its desire to claim to know things of which it could not have experience (e.g., God) by grounding knowledge in experience. But he was also among those who realized how much Kant had compromised in order to secure reason, most significantly by giving up any claim on the world in its immediacy, thereby consigning humans to living in a world of representations. Guided by Johann Gottlieb Fichte in their early years, a young Schelling and others sought to overcome the divide between human knowledge and the world, to seek out what came to be called the absolute, the infinite, the unconditioned, and, in Schelling's rendition, nature. This was not just brute nature but nature as the site of experiences of the sensible as well as the realm of the supersensible, the transcendental, and the ideational.

While still grounded in Kant's rigorous method and architectonic of thought, Schelling was involved in a wide range of experimentation on how to think of nature beyond the mechanical laws that undergird it or as a domain of necessity to being a source of freedom and creativity. These experimentations constituted the basis of his *Naturphilosophie*. Two aspects of his *Ideas for a Philosophy of Nature* ([1797] 1988) stand out to his first-time readers. First, Schelling's view of nature encompassed attention to inanimate matter and its motion, the stars above and their movement, biological organisms and their rhythm, and thought and its striving, including that of human consciousness. I was struck by his refusal to adjudicate in favor of either matter or mind, keeping both in the picture and being attentive to the contours of their particular movements in time and space. I took this to be a commitment to both empiricism and idealism and their mutual imbrication, something I had also found in the chaura context. Second, I was struck by the fact that Schelling's arguments did not progress by means of the dialectic alone in which a contradiction between two premises produces a third that supersedes the two while still retaining them, a method that has come to be identified with Hegel, to whom the young Schelling was also very committed. Rather, Schelling discerned oscillation between two polarities within each scale of the organization of matter, with the dynamic of movement produced by intensification/contraction and expansion. This picture of existence as one of oscillation between two extremes also sat well with the conditionalities that informed chaura lives.

A closer reading of Schelling's work, specifically *System of Transcendental Idealism* ([1800] 1993) and *First Outline of a System of the Philosophy of Nature* ([1799] 2004), yielded nature as pure activity, excessive productivity, and dynamic movement. It did not see a meaningful difference between inorganic matter and organic matter, producing one of the most profound meditations on matter as constituted of forces. It also did not see a difference in kind between consciousness and forces, with mind in a continuum with matter. And while nature could be determinate and directional, presenting itself as finite forms and intelligible laws to human consciousness, given its dynamism it was also oriented toward delirious excess, indeterminacy, and the dissolution of forms. The human comprehension of nature meant recognizing that nature was within one, in one's consciousness and unconscious, that it required imagination and intuition to grasp human participation in nature, and that as a species we were bound to the same indeterminacy as inorganic matter.

Some of the questions that I saw Schelling pose over the course of his life and that I have privileged in this book are, in a world in which nature is present as a force, What are the possible relations between force and materiality, between mind and matter? How is nature both within and without us, expressed through our activity and passivity? Given that we cannot cognize nature except as appearances, as Kant laid down, what are other modes of accessing nature in its immediacy, such as through our imagination? How is nature complicit in human projects, for good or evil? Finally, How does our mythology give us a sense of the mythological as actual, a sense of our historical evolution as peoples and a sense of nature?

The first of Schelling's questions—that is, regarding the possible relations between mind and matter—became the basis to suggest a different relation between the chauras and the char land. While this relation is primarily understood as property and patrimony, I claim that it could also be understood as a join between matter and mind, imaginable through chauras extending the life of the land by means of strife and striving. The second of Schelling's explorations in conversation with the esoteric philosophy of Jacob Boehme asked how nature is within us. This led me to explore whether gaps in chaura narratives on what they did and why they did those activities in the face of erosion of their lands may be seen as nature within them, in their constitution and mental makeup. Next I studied chaura attempts to mobilize elections in many different ways, including but not restricted to ensuring elections for villages that no longer existed to secure the future existence of the villages. Their playing and replaying in their mind's eye the events of erosion that led the villages to disappear, and their future-oriented projections as to how elections would bring together villagers and possibly ensure the future of villages, served as an important instance of chauras' reflection on their place within the workings of nature. I argue that instance could serve as an empirical example of the intuition that Schelling and his mentor Johann Wolfgang von Goethe saw to be the means to access nature in its immediacy. While Schelling's understanding of the unconditioned as holding the possibility of evil within itself was his response to the question of theodicy—that is, How are human acts of evil understandable within a world in which God is present?—chauras' acts of forgetting Hindus with whom they had previously coexisted modulated

Schelling's high theological mode by suggesting that such acts may not be so much evil as just in the nature of things, of river branches, lands, and human memory to fall into ruin. They are more begetting of injustice than evil. Finally, while Schelling's late philosophy of mythology was interested in its tautegorical nature, as presencing God's revelation in its immediacy, within the chaura context this perspective on mythology was extended to understand how nature creates culture for which mythology is an important means of expressing itself, how small acts carry mythological weight and insight into the workings of the world.

Far from a heavy-handed application of abstract philosophy to lived experience, I take the relationship between Schelling and chaura existence to be one of conviviality—of which Andrew Brandel says, "[It is] life with those who offer not only competing answers to our questions but also competing questions" (2016, 324). As Stephan Palmié (2018) and Veena Das (2020) elaborate, our decision as anthropologists whether to focus on self-enclosed ontologies or contaminated and crisscrossed thinking also reflects our understanding of whether we imagine we have a future in one another's languages. This is not a fantasy of perfect commensuration and translatability but a vision of coexistence in the same world or adjacency to one another's worlds. I claim an adjacency between Schelling's theologically inflected understanding of nature and the chauras' God-saturated naturalism and stage a conversation between the two over the course of the book.

Structure of the Book

Although I continue to return to my field site, this book is based on fieldwork largely conducted between 2011 and 2017. As mentioned earlier, I selected three villages to study. I carried out classical research consisting of mapping village neighborhoods, including landownership; carrying out surveys of 10 percent of the households in each village; undertaking repeat interviews and participant observations in everyday settings and at all the major events that arose during the times of my fieldwork; doing family genealogies; tracking the agricultural seasons; learning about health and illness; inquiring after domestic and wild animals; learning about the intersections of the villages, households, and individual lives with regional and national politics and economy; and so on. I also innovated on techniques, such as producing movement maps of individuals in the mainland and those in the char to have the basis for comparison across a cross section of the landscape. I created archives of newspaper cuttings, land-related legal cases, NGO reports, religious manuals, farmers' almanacs, story collections, and CD/DVDs of popular musical events, such as *pala gaans* (narrative songs). I repeated portions of the research during different seasons to see what changed from season to season, and I also repeated the research annually to see what changed from year to year. As villages eroded and people dispersed, I had to do considerable sleuth work to locate and follow up with those being studied. The continual changeability and mobility of my study population, combined with the fact that after I was done with my research leave I could only return during times when my university was not in session, give a certain episodic quality to my ethnography.

Chapter 1, "Moving Lands in the Skein of Property and Kin Relations," outlines the fights produced of a long history of survey, settlement, and disputation over char lands. The chapter seeks to give a sense of how law, lands, and lives are imbricated, making for a fraught sociality. At the same time it shows how this imbrication also produces a kinship assemblage that extends the arc of land beyond the occasion of its erosion. The chapter shows how expectations, desires, and anticipation modulate the vicissitudes of land.

In chapter 2, "History and Morality between Floods and Erosion," I explore the narratives of the chaura experience of erosion to show how they think of themselves as literally within the river, as entrained by the river in the same way as the river entrains sediment and vegetation. This contrasts with their experience of floods, which shows them to be within the time of the nation-state, and this experience of entraining hints at how the river may be in one's unconscious.

In chapter 3, "Elections on Sandbars and the Remembered Village," I trace how chauras displaced by the erosion of their villages return to vote politicians into office for their lost villages. The energy those living in chars invest in electioneering for villages that no longer exist points not just to self-interest in keeping one's territory alive within national maps but also to efforts to intuit one's participation in the hanging together of matter. Imagination is opposed to intelligence as a way to know nature.

In chapter 4, "Decay of the River and of Memory," we see an instance of chaura lives in which the Muslim chauras enact the erosion of villages and the degeneration of river pathways as the loss of memory of a shared existence with Hindus who lived in the villages with them until the recent past. The insight that evil is less metaphysical and more in the nature of things to fall into ruin, to be unjust, helps to explore chaura (in)action from the perspective of human intentionality *and* the dissolutive tendencies of nature.

In chapter 5, "Death of Children and the Eruption of Myths," I explore how the fading mythology from a past of shared existence with Hindus retains salience beyond that past. The myths of the goddess Ganga Devi and the living prophet Khwaja Khijir linger to express how Muslim chaura women experience and understand the loss of their children to watery deaths and to show how nature thereby enters the mythic and founds culture through women's dreams and discourses.

A short epilogue, "The Chars in Recent Years," tracks my last visit to the field site to update myself on the changes to the milieu. It remains attentive to the aspect of nature within the stories the char dwellers tell of themselves even as they speak of new horizons.

Approaching the char, it was hard to tell where the river ended and the shore began. Photo by author.

The fields around Dokhin Teguri
were lush and the village hemmed
in by trees. Photo by author.

If one looked more closely, the fields
were pocked with large craters
with their sides collapsing inward.
Photo by author.

The approach to Rihayi Kawliya gave
notice that the work of erosion
was more immediate and catastrophic
than flooding. Photo by author.

The denizens of Rihayi Kawliya had taken to living on new char land amid catkin grass. Photo by author.

Those newly transplanted to Kuwait
Para in Boro Gorjan after the land
of their villages had eroded. Photo by
author.

The houses in Kuwait Para backed onto a branch of the river. Planting *komli* (a shrubby plant) provided some modicum of physical and psychic separation from the waters. Photo by author.

Moving Lands in the Skein of Property and Kin Relations

Land and Property, Matter and Mind

WHEN CHAR LAND FIRST APPEARED in the Jamuna waters, the chauras living close by approached it cautiously. After all, the land might have appeared temporarily, and there was no point in investing too much hope in it until at least the seasons changed, or a year or two passed. At other times, however, land didn't even break through the water surface before chauras were busy planting rice seeds to grow into saplings for transplantation. The hope was to fully utilize the land for the short time it was around. The difference in approach was contingent on many factors such as the season when the land came up and the quality of the soil. If it was the rainy season and the accreted land was sandy, then the land might not last out the season. If, however, it was the rainy season and the land was loamy, consisting of a combination of sand and sediment, then it was prime for a short stint of rice or lentil cultivation. But even if the land was sandy and not ready for cultivation, if it appeared during the winter season and was somewhat stable, chauras planted catkin grass (*kayisha* or *kashbon*) that grew to tall heights and helped make the land firmer and fertile. Most crucially, however, the difference in char dwellers' approach to land was mediated by whether the owners of the land were relatively lax in their surveillance of their property, allowing the land-less to take advantage of it in the early days of its emergence.

An aspect of the char landscape that emerges from these practices: that cultivation began from the earliest possible moment of the emergence of land, either to take advantage of its fecundity or to bind its soil to protect it from erosion. Land informally operated as public commons for a period of time before being reinstated as someone's private property. These practices could be undertaken even when there was no land in sight. In other words, it took physical labor to keep char land intact and fertile. But equally important to keeping it intact was the work invested in the legal fiction of the land as property. This orientation toward land as property suggested the availability of the matter that made up land to be taken up not only through its quality of soil or the commingling of one's physical labor with it, but also through one's mental labor.[1]

While it is beyond the scope of this book to provide an adequate review of the concept of property within political theory and anthropology, suffice it to say that we are used to thinking of property as that with which human labor has been intermingled, a Lockean formulation of possessive individualism (MacPherson 1962). Georg Wilhelm Friedrich Hegel separated the person from the possessed thing by instituting a relation of right between the two (Hann 1998), while Karl Marx ([1930] 1988) and Friedrich Engels ([1888] 2010) contributed to our understanding of private property as necessarily alienated within the wider context of capitalism in which a person's labor, itself transformed into property, produced things as objects on behalf of capital.[2] Given this history of understanding property, it may be surprising that I propose that chaura efforts to maintain the status of eroded and accreted land as property be seen within a different analytic than one of owner and object, rights, or even social relations. Rather, we may see land-as-property as mind straining to be matter, as matter seeking to be present in absentia. This gesture did not hold for property in general but specifically for the chaura, a group of people who were put upon by the landscape on which they lived and whose mental labor involved appropriating standing legal frameworks in the service of extending land. In making this argument I draw inspiration from Schelling's *Naturphilosophie* when he writes, "The human mind was early led to the idea of a self-organizing matter, and because organization is conceivable only in relation to a mind, to an original union of mind and matter in these things" (Schelling [1797] 1998, 35). I read these words as suggesting that because mind arises out of dynamic matter, it lends its capacities to matter to enable matter's organization to express itself. Of course, anthropology will rightly insist that any such relationship between mind and matter is mediated by the social, and this mediation becomes amply clear in chauras' efforts to extend the lives of their lands-as-property by means of their lives and interrelations.

Before we consider how chaura lives extended lands through living and relating to one another, it is necessary to first ask what kind of a property regime allowed for the continuity of the rights of ownership, or more specifically the record of rights (khatiyan) and title deed of ownership (dalil), in the absence of actual land.[3] How was this property regime related to chaura lives? We have to entertain the possibility that it wasn't just the land regime that produced chaura sociality, but that chauras took lands, or more specifically fights over them, as the occasion for their interrelating, with material consequences for the future of the lands. Only then will we be in a position to answer how, by plying their own lives and relations, the chauras were able to perpetuate char land, to serve as its extension.

The British colonial administration undertook the last cadastral survey (CS) in this area, by which legal holdings were stitched to actual physical boundaries and put in relation to neighboring titles. The CS, as it is commonly called, was carried out between 1888 and 1940 under the purview of the Bengal Tenancy Act of 1885. No sooner did the CS come to an end than the British began another Revisional Survey (RS), which noted the names of the owner and of the possessor, who very often were not the same person.[4] After the British quit India in 1947 and India was partitioned into India and Pakistan, a Settlement Attestation (SA) was undertaken by the government of Pakistan between 1956 and 1962 under the auspices of the State Acquisition and Tenancy Act of 1950. The SA was based on self-reporting by land possessors, not on field surveys and was therefore riddled with deceptions and errors. While Bangladesh has undertaken several mutation surveys to record transfers of property, its own cadastral survey (the Bangladesh Survey [BS]), which has been ongoing since 1984, is still incomplete and has yet to be extended to the parts in Sirajganj where I lived and worked. Depending on the years of emergence of the char lands in this area, they might have CS, RS, or SA records, or none at all.

During my fieldwork between 2011 and 2017, I found chauras holding many kinds of paper documents attesting to their private ownership of land: copies of records of rights from the CS or the RS and even the SA settlement maps (either handed down from their forefathers or acquired through special favor or bribes from the *zila*, or district-level Office of Land Records and Surveys); actual title deeds, although these were rarer; registration forms of transfers of property (obtained from the *upazila*, or subdistrict Land Registration Office); and handwritten transfer notes. These were kept as safe as possible, wrapped in plastic bags, sometimes even laminated, and put in metal trunks with large locks to protect them from water, mice, and insects more than from other humans. No one had any recent records from the BS begun in 1984, although they were all waiting for this survey to come to these parts. In the meantime, char land came and went. Chauras transacted in lands that didn't exist and farmed or fought over those that did.

As I indicated in the introduction, the British were unsuccessful, or rather strategic, in their efforts to survey and settle the property determinations of char lands through what were called surveys of water bodies in the daria region, an active floodplain area of the Ganges River. They considered it a bit like "drawing lines on the sand" and specifically avoided lands of a "fluctuating nature" (Hill 1997, 47). Colonial prevarication toward chars produced a plethora of acts, laws, and legal precedents that were inherited by postcolonial East Pakistan, later Bangladesh. These have enabled and enforced the legal fiction of land as property even when the land is physically absent. Many activist scholars (Barkat, Zaman, and Raihan 2001) consider this legal heritage to be at the root of the violence over possessing unclaimed lands that characterized char sociality and strongly advocate that land laws be radically reformed, property determinations be made in favor of the landless poor, and that the BS be extended to this area to secure legal holdings "once and for all."

The commonsensical understanding here is that sorting out land laws and property holdings will straighten out chauras' social relations and reduce the violence among them. But it is also noteworthy that the property regime I sketched earlier allowed for ownership in the absence of land. This sense of ownership kept chauras in and around absent lands. It also allowed them to have something to transact when land went under. It exerted a pull on a wider network of relations than the individual owner/possessor alone, ensuring a collective movement of people and their settlement and resettlement on moving lands. This is not to say that I oppose land reform—far from it. I am instead making a separate claim. It is not sufficient to look at land laws, ownership patterns, or property papers as the sole basis of char-based sociality and violence. While land generated its distinct interests, whether in its market value, in its crop yield, as guarantor of power and status over others, or as security or collateral, more often than not what was at stake in char-based fights over land was who was kin to whom or who was implicated in whose existence. In other words, it was kinship determinations that stood to extend the life of the land beyond its presence through legal title.

Marshall Sahlins's idea of kinship is useful here. As he writes: "The specific quality of kinship, I argue, is 'mutuality of being': kinfolk are persons who participate intrinsically in each other's existence; they are members of one another" (2013, ix). With this statement, he aims to bridge the gap between the old kinship, which was heavily classificatory, functionalist, and normative (see Graburn 1971), and newer kinship studies, which privilege relationality as the mode of making kin and are actively averse to normativity (see Franklin and McKinnon 2001). While Sahlins feels liberated by the new kinship to be able to pitch a more encompassing analysis of kinship, he is nonetheless careful to explicate that flexibility and contingency in kinship relations did not obviate shared understandings, that relations carried moral expectations and jural obligations.[5] Thus, in emphasizing mutuality of being, he called for a balance between principles and relations, or the normative order and the negotiated reality. In the case of the chaura, that would mean the balance between the Bengali Muslim order of kinship and what was possible and privileged under chaura conditions of life.[6] I would further suggest that just as the kinship order was not explicit and only came into view at specific moments of relating, mutuality also was not a given. It took work, specifically the work of fighting in the case of the chauras, to clarify the existence and extent of this mutuality. Fights over land are privileged in my account over other kinds of fights because many of these other fights involved intimates and were folded within fights over land, either lending a spark to land-based conflagration or receiving new energies from them.

In this chapter, I examine the skein of Bengali Muslim kinship activated by land-related conflicts in the chars as a way to explore the thickness of relationality that made life in chars possible in the absence of state services and institutionalized safety nets and that gave char-based kinship its specific texture, with the threats of force, violence, and disappointments imbricated in it. What interests me is how a focus on land, manifested in fights, helped one to assemble one's kinship relations. These kin relations were crucial to creating patches between the different parts of lives regularly disrupted by erosion. But fights over land also animated certain lines of connection to specific kin such that land, even land in absentia, came to have a caretaker, someone who kept watch at that part of the river where the land once was, who ensured that government efforts to drop soil or scour the bottom of the river were encouraged or foiled in the interest of the land, and who was there to greet the land when it emerged. Thus, in addition to participating in the legal fiction of land as property, chauras participated in tending and extending the life of lands through their care and attention to it, brought into focus by fights over the land. At the same time, this fixity of faith in the existence or imminent arrival of land produced a blindness to the nature of the terrain, such that every time land finally fell into the water through erosion, chauras were jolted, shocked that such a thing could happen. I am interested in how a focus of attention specifically to land and kin suppressed focus on the moving nature of land, rendering the surety of erosion into an unknown, unlikely and unimaginable. At the end of the chapter we see how char land as property produced relations and attention, which in turn extended the duration of land but only for so long. Nonetheless, the coil of land as property, the fights over land, the determination of kinship relations, and the care and attention to land speak to the imbrication of matter and mind without collapsing the distinction between the two.

The Swell of People at Phulhara
(Village of Lost Flowers)

When I first visited Phulhara in 2010 with the head of MMS through whose generosity I later came to be living in the Jamuna chars in Sirajganj district, it was summertime, and we had to take a boat to the village. This village obviously was the pride of the NGO. As I walked around and visited a local school, I saw a well-disciplined, one could even say a model, Bangladeshi village of that decade. The homes were clustered on upraised land in a picture of self-reliant villages as fostered by the state and NGOs concerned to keep landless rural people from migrating to city slums (Ali 2010). The children sang nationalist songs and read from texts in proper Bengali that excoriated the social ills of rural life, such as young girls being given into marriage before their time and poor farmers being driven to bankruptcy because of their debts to rapacious moneylenders. Yet when I returned a year later and went to visit on my own, it felt as though what I had seen earlier had been a carefully constructed front put up for visitors and donors and the second visit revealed the seamy reality behind it. Over time I would realize that both faces were true to this place, that a desired assimilation of an egalitarian, nationalistic NGO ethos existed alongside a commitment to the sense that everyone had an appropriate place, with some occupying better places than others. Such thinking was also exercised toward one's own children such that it was clear from early on in the lives of children which one was smart and deserved education, and which one was "brain short" and should be expected to care for domestic animals.

On the occasion of my second visit to Phulhara in 2011, I was visiting a few of my acquaintances when a swell of people passed the courtyard of the house where I was standing. There were men, women, and even children, almost all brandishing makeshift weapons: a broom, a stick, or a brick. I even sighted a boti or two, a curved blade affixed to a solid base that women used for cutting up food. It was a terrifying sight, and the villagers who quickly gathered in their courtyards to watch the crowd had worried looks on their faces, although they also seemed to be encouraging the crowd along. A woman carrying a stick, one of the "beneficiaries" of the many NGO-sponsored projects in the village, called out to Shohidul, my research aide who also worked for the NGO, to jokingly ask if he wasn't glad that she was *out of the home*, an index used by NGOs to measure women's empowerment.[7]

The people were heading to a designated place in the middle of the paddy fields to meet an opposing crowd for a *kayija*, or fight in the old style. When I asked my acquaintances in Phulhara what exactly was going to transpire at the site, a few bashful young men performed for me the *lathi khela*, the martial dance involving long bamboo sticks upon sticks, associated with northern Bangladesh. I had seen renditions of this dance on television as a performance of the culture of Bangladesh but until this point it hadn't dawned on me that this was an aestheticization of actual physical confrontation and acts of violence.[8]

Of the two groups confronting each other that day, the Phulhara group represented a village big man by the name of Yusuf Member. The opposing group from Chaluhara (the Village of Lost Wits) represented another big man named Shukkur Member. Both had once been members of the union *parishad*, or council, the lowest tier of state government. Although they were no longer members, but not for lack of trying, their titles had become a permanent part of their names as with all representatives of the union councils, causing the titles *chairman* and *member* to proliferate across the rural landscape. These two, Yusuf and Shukkur, were fighting over land that Shukkur currently farmed. Because the land was closer to Phulhara village than to Chaluhara, Yusuf believed he had rights of proximity, and he had made his feelings clear by having Shukkur ambushed and beaten, landing him in the hospital. The people of Chaluhara coming to fight those from Phulhara were enraged at the treatment of Shukkur.

In an island where one is hard-pressed to see any of the usual signs of government, even a loitering police officer or a cycling postman, it was surprising that the kayija was stopped by a group of policemen who had come from the mainland obviously for the purpose of preventing the fight. People speculated that either of the two sides could have called in the police. It may be that while there was a desire for the spectacle of groups threatening each other, no one had the stomach for violence. The almost-fight raised the question of the property regime that would allow for its possibility.

Khas in the Weave of Land Laws

As Ranajit Guha (1996) has observed, colonial land laws and administration in Bengal began with great aplomb with the Permanent Settlement of Bengal of 1793. The visionaries in favor of generating a proprietary class moved by the spirit of improvement were soon reduced to witnessing the growth of a rentier class of zamindars dependent on revenues generated by the raiyats or under-raiyats, peasant cultivators, working the land. The subsequent Bengal Alluvium and Diluvium Regulation of 1825 allowed the British to go around their own settlement (Hill 1997; Soeftestad 2000; I. Iqbal 2010). It gave them the opportunity to develop direct relations with the raiyats cultivating riparian sandbars and coastal accretions by classifying these as khas, or government-owned land, and therefore directly taxable by the state, thus circumventing the zamindars. Zamindars challenged state access to these lands, which they saw as their own, eroded and reconstituted elsewhere. The colonial state accepted their claims if the new land had accreted to property already held by a zamindar or was fordable during dry periods of the year, but otherwise it claimed freestanding char land as its own.

As the rights of zamindars eroded over the course of the nineteenth century with growing state recognition of the occupancy rights of the raiyats, legislation veered back and forth between giving more strength to these rights and determinedly holding on to the possibility of declaring char land to be khas. The Bengal Rent Act of 1859 accepted tenancy rights to new accretions provided the tenant paid more rent if the new land accreted to his property and less if his property decreased, but the Bengal Tenancy Act of

1885 warned that the tenant would lose rights over his property—that is, it would revert to being khas—if he accepted a reduction in rent. The Tenancy Act of 1938 automatically reduced rent at the loss of property, giving the tenant the right to retain the land if it re-formed in situ within twenty years. The East Bengal State Acquisition and Tenancy Act of 1950, drafted under the colonial regime but put into effect during Pakistan times (1947–72), was the last blow to the zamindari system, which was formally abolished in 1956. This act granted full occupancy rights to raiyats but was later qualified by a presidential order in 1972 in newly formed Bangladesh in which all land formed in situ or as new accretion was to be automatically considered khas whether or not it re-formed within twenty years. This order was amended again in 1994 to allow tenants, now officially landowners, to retain rights over their land if it re-formed within thirty years of its dissolution, adding an extra ten years in which to hope for the return of one's land, but set sixty bighas (approx. thirty-seven acres) as the maximum amount of this land that could be reclaimed beyond which it would be considered khas land.

From this sketch of land laws and articles within laws relating to char lands (see also Sirdar 1999), it is readily apparent that the enduring ambiguity lay not in the arbitrary nature of legal pronouncements but in the flexibility retained by the state, whether it be colonial, Pakistani, or now Bangladeshi, to classify land as khas. Moreover, these were not the only laws by which land could be deemed khas. The term khas traditionally refers to wasteland in the state's care, such as swamps and jungles; land used for public purposes, such as markets and riverbanks; and sites of natural resources, such as sand bodies and stone quarries. As mentioned previously, it was by a clever sleight of hand that chars were designated as khas by the colonial government, to allow them to grow their revenue base beyond zamindars. But even those lands that had fully established rights of ownership over them retained the capacity to become khas if the char appeared a year too late or a yard too long in the course of accretion, as is clear from the preceding history of land-related legislation.

Adding to its authority to declare land as khas, in 1947 the East Pakistani government took over as khas all land abandoned by Hindus leaving East Bengal for India during partition. When the zamindari system was abolished in 1956, the state bought all land from ex-zamindars in excess of the then-current ceiling of 365 bighas (121 acres), declaring it khas land. And, in 1965, the state put into effect the Enemy Property Act (later renamed the Vested Properties Act) during the India-Pakistan war by which to confiscate and declare as khas any property left behind by Hindus leaving for India during this period or those who had been charged with sedition. This act, only repealed in 2001, has been used to confiscate the land of many, including that of Muslims but predominantly that belonging to Hindus (Guhathakurta 2012).

The khas classification has thus acquired a certain spongy quality that allows it to envelop and absorb the diverse properties of land. This ability to clear land of any historical associations or memories was most prominent in the informal use of the term khas by char dwellers to designate lands that were mapped in the first CS (which was carried out between 1888 and 1940 and produced the settlement records) but were nowhere in sight, or were underwater, during the state acquisition or SA surveys conducted between 1956 and 1962. When such land reemerged after long periods of time underwater, the mutation of property, by which is meant change of ownership, was either unclear or hard to establish such that the truth of its ownership was determined by dramatic fighting over it. The informal characterization of such land as khas opened it up to char dwellers working out the specificities of the law on the ground, but for this to happen khas first had to acquire its expansive quality by the particular weave of the law and the markets for such land.[9]

The Market in Lost and Found Lands

While for the government khas had a clear meaning as land belonging to it, in the chars it meant several things. It could simply be wasteland, useless for cultivation, designated as such in colonial settlement records. It could be land confiscated from zamindars after the abolishment of the zamindari system in 1965. It could be land confiscated from Hindus who had emigrated to India or enemies of the state who had simply absconded in the early 1970s. It could be alluvial land that emerged beyond the thirty-year mark from some distant past but attached to one's existing land such that one could claim it as a growth of one's property. Or it could be land with ambiguous rights of lease, such as land customarily associated with zamindars but now in the state's purview, with both parties retaining the right to give permission to till, and each permission being likely to come into conflict with the other. One mode of laying claims on this land was by physical fights of the kind described earlier. Another was by creating a gray market in the buying and selling of such land to give it a patina of legality. Qadir, a land dealer, provided me with a description of the market for such land, which was only a small part of a larger market in which people transacted in lost lands to which they claimed to have legal titles.

The larger market involved people with titles to land from the first time it was surveyed, marked within settlement records, and given a plot number (*dag*) and record of rights (khatiyan) maintained by the Office of Land Records and Surveys under the Ministry of Land. These might date to the administrative regimes of either the British Raj, East Pakistan, or Bangladesh. Though the land might no longer exist, owners who sought to transfer property were able to do so at the subdistrict subregistry office maintained by the director of land registration under a different ministry, that of law rather than land, which overlooked the absence of the physical land being transacted. The shared understanding was that once a new survey was done of the lands, providing they had reconstituted, the new owners would be recorded in the updated survey records and registry. But the new owners would not be issued new titles until the survey was undertaken. Until that time they had only slips of paper showing that they had put in applications for title changes and paid the necessary property transfer taxes. These slips came to function as de facto title deeds to be bought and sold, with each transaction and change of hands involving further slips of paper from the registry offices. People carried around deeds that named old patriarchs and matriarchs as owners of their land (although they might be several generations younger than these original owners or may have constituted entirely separate family lines), along with registration receipts for each subsequent transaction. It was important for any new owner to acquire the entire bundle of papers to be able to show a clear and undisputable line of property transfer, or so the hope went. Since a new survey had yet to happen and the rules were not in place, much of this trade was done in the hope that these transactions would be granted state recognition.

In 2014, the subregistry offices stopped accepting applications and payments for changes in ownership of land. This led to another layer of documentation, which was sometimes no more than a handwritten note with thumbprints and an official-looking stamp or a signature from an elected official claiming there had been a further transaction of the said land. This system was undoubtedly rife with forgery as evidenced by the fact that some people had only photocopies of original title deeds and registry receipts, and I knew many who were involved in litigation to authenticate their claims on land, seeking to get a "degree" in their favor from the courts, which was another way to secure one's claims on land.[10] Thus, while waiting for the

BS to come, chauras tried to fix their claims as best they could through the previously mentioned chains of documents; quasi-legal modes of sales, transfer, and registration; *and* physical fights.

Transactions of khas land within this quasi-legal space did not result in the production of fraudulent title deeds, as Qadir, the land dealer, explained to me. The Office of Land Records could easily detect such frauds by checking the original records. Anyone wishing to make a lasting claim on khas land, particularly land known to have once belonged to Hindus who had left the area, had to produce owners, very often fraudulent ones, who had to claim a clear genealogical relationship to the original owner in order to be furnished with authenticated copies of previously held title deeds from the Office of Land Records. This person then had to appear before the sub-registry of the Office of Land Registration for the transfer of property and stand up to the scrutiny of the local revenue inspector (*tehsildar*) and the assistant commissioner for land. Only after new title deeds had been issued was the change to be recorded in the original records maintained by the Office of Land Records. I suspected that this last part did not usually happen, leaving the original records intact while the plethora of transactions I have sketched here accrued without leaving any tracks in the official records.

The efforts taken to find just the right person to claim to be the owner of khas land were astonishing. Qadir was well known in these parts for tracking down someone who could pass for the owner's lawful inheritor. To this end he traveled far and wide in Bangladesh and sometimes to neighboring India to find a person who could supply names of forefathers going back several generations, provide vivid descriptions of the owners of adjoining plots, and offer explanations for why they no longer had any need for this "useless piece of property" and felt the time had come to let it pass on to "poor people who could make better use of it." Only such a performance enabled the issuance of title deeds to enable the licit sale of illicitly claimed land.

To now return to the fight that was ongoing in Phulhara between Shukkur Member and Yusuf Member, each of whom had produced a Hindu from whom they claimed to have bought the land: How, then, were their conflicting claims on the same plot of land resolved? Given the sensitivity of the issue and the degree of violence involved, I was advised not to question them directly at that moment; hence, most of my narration is in the form of reported speech. As it was, I would only find out the outcome a year or two after that original event of the almost-kayija, when the dispute had been fully resolved. I heard that there had been a local *shalish*, or dispute settlement, in which both men had presented their cases, including their title deeds, showing that they had bought the land through the licit-illicit practices I have described. The final decision had been in Shukkur's favor because he had perpetrated the fraud earlier than Yusuf, and this was widely known. A wide acknowledgment of this led to questions by the shalish as to why Yusuf would follow up Shukkur's fraud with a fraud of his own unless he sought to create mischief. In my observation of numerous shalishes, I had noted that the interest of a shalish was the perpetuation of normative order and political stability within rural life rather than attending to the particulars of legality or to individual rights.[11] Therefore, any intimation that Yusuf had mischief in mind rather than a genuine desire to claim and cultivate land was enough to swing the decision against him. And so the conflict came to an end.

Kinship in the Midst

It should not be a surprise that many of the men involved in kayijas came from a loose group of *matbors*, or headmen/influential people, who first showed up the moment land emerged from the river to stake claims on it but who also put to order the people who followed behind. They were the ones who enabled collective cultivation as the land became arable. Later they had to abdicate claims on the land as they were followed by more settlers accompanied by *amins*, or unofficial land surveyors, who pointed out to whom the land rightfully belonged according to settlement records. But these matbors still managed to squeeze out a bit of land in concession for the initial leadership they provided. They usually stayed around the village in diverse capacities, slowly turning from matbors in the capacity of headsmen to matbors meaning busybodies, and they waited for the land to erode to guide people to other chars or for new land to emerge so they could jump on it.

This was the origin story I was told for each village in the char, although usually people were hard-pressed to name the founding matbors of their own village. Villages were not in the control of single leaders who might arrogate to themselves the right to monopolize all resources. This was no longer a possibility with the state regularly introducing new leadership through the electoral process and political parties organizing at the village level (Siddiqui 2005; Lewis and Hossain 2008). Conversely, landownership was not the definitive means to a position of power, if it ever was, with wealth in people counting for more than wealth in land (Guyer 1996; Mendelsohn

1993; M. H. Khan 2004). And it was unlikely that land being fought over was for farming as Shukkur Member had plenty of land lying fallow near his own household that he was unable to farm for the lack of people willing to provide the necessary labor. It was a common complaint among landowners that NGO uplift of the poor had left the landowners without access to cheap or bonded labor to work their lands.[12] The owning and farming of land in Phulhara, far from Shukkur's household, almost in someone else's backyard, had to have other stakes than the obvious ones of cultivation.

As anthropologists have long stated, the rights over a thing or property relations are almost always about relations among people (see notes 1 and 2). They further say that the right isn't only over a thing; it is also the right to be a certain person or to lay claims on other persons (Strathern 2006). An intimation that it was relations between people at stake in the almost-fight in Phulhara first appeared in the days following the fight. In the courtyard of the MMS office where I made my home, I witnessed a moving interchange between two elderly women, who turned out to be the sisters of Yusuf and Shukkur. They were both beside themselves in sorrow over the beating Shukkur had received at the hands of Yusuf's henchmen. Soneka *khala* (aunt), Shukkur's elder sister and an informant with whom I was very close, had come running to the courtyard when she heard Yusuf's sister had come to take out a loan from the credit office of MMS. The two clutched each other and cried as Soneka pointed to parts of her own body as she described where Shukkur had been beaten. The interchange was strange to say the least because not one word was uttered, but strangled sounds issued from both to indicate the pain that was received by Shukkur and felt by the two women in unison.

That was the moment I learned that Shukkur's mother and Yusuf's mother were sisters. In Bengali Muslim kinship, in which exogamy is the rule, marriage into a *bongsho*, or lineage, means exile from one's father's lineage and absorption into one's husband's (Inden and Nicholas 2005). Two sisters living side by side indexed merely two separate lineages side by side. Otherwise, there were no clear rules or obligations between the two lineages, and their children were not bound to one another in any way other than that of affection. Unless one lineage was stronger than the other, the relationship between matrilateral parallel cousins was that of equality. Such a relationship existed between Shukkur and Yusuf. While Yusuf's village had suffered little erosion, Shukkur's village, Chaluhara, had broken and reconstituted many times in the life of this man in his sixties. He no longer made his home in Chaluhara, preferring to live in the more stable village of Dokhin Teguri at the very edge of Chaluhara from which he could keep an eye on developments within his village. And he might not have farmed his land in Chaluhara because he did not trust it to last long. Instead, he might have invested in the elaborate fraud necessary to acquire and farm the land in Phulhara at a distance from his existing household to ensure that he had a piece of land to move to if his present home eroded. Shukkur might even have chosen this land near Phulhara trusting that his cousin would keep a watchful eye on it for him, as Yusuf was also fairly powerful in his own right. For Yusuf to turn against him by making bids on the land, having Shukkur beaten and then subjected to the expenses of a shalish, in which each presiding figure drew hefty fees from both sides, indicated a spurning of the affectionate relationship between the two.

We may never know what made Yusuf act this way. This was a period in his life when it seemed as though his power and authority in his own village of Phulhara were ebbing. He was already involved in contestations not just with other big men but with poor women in his area who were openly accusing him of pilfering from the welfare benefits they were due for their labor in cutting earth to make roads in his village. What became apparent was that within a kinship system that primarily privileged *shoreek* (intimate relations, usually bound by blood) and *kutum* (relations through marriage), here was a distinctly new emphasis on a part of the kinship network that had previously not merited much attention, that between matrilineal parallel cousins, embodied by the tempestuous relationship between the two male cousins and the loving one between their sisters.

The Mother of All Fights at Doshayiga, or the Village of Ten Paces

That kinship was at the heart of land-related fights was indicated by the mother of all fights in the village of Doshayiga over the 1970s and 1980s. I first heard the story of Thandu and Nannu from Shohidul, now in his early thirties, who had witnessed the spectacular fight between the two men when he was ten years old and was reminded of it while watching the crowds in Phulhara. Shohidul recounted how on a winter's night in 1980, in the Bengali month of Agrahayan, Thandu and his men stayed up cutting the aman rice paddy from a patch of land in the village of Doshayiga, or the Village of

Ten Paces, adjoining Thandu's village of Bishtipur, or the Village of Rain, in the district of Sirajganj not far from the village where I made my home. This land had emerged only a few years prior to this event. By early dawn, the fields were cleared and the crops gathered for transport. The men washed their tired faces with the cool, brown waters of the Jamuna; straightened up; rubbed mud on their bodies to show their determination to fight; and, picking up lathis, the long bamboo sticks ending in sharp metal spokes, braced themselves to face Nannu's army. I imagine that Nannu and his men walked across villages and fields to the designated site of the fight as this land was still qayem, or mainland, at the time of the kayija. The fight involved stick on stick, knife on knife, ax on the exposed parts of bodies, and the occasional brick. The fighting continued, interspersed with an occasional shout or grunt as if the men were furiously at work rather than in pitched battle with one another, until the point at which Thandu was gored.

Shohidul described how Thandu, with his coils of black oiled hair, flashing eyes, and fierce demeanor and sporting a knife contrived of many blades, was the one to watch. When the fatal blow struck him, a wave of weakness rippled through the line of defense as several other men also received severe wounds. At the sight of blood, Nannu's men dropped their weapons and ran. Although blood was exactly what was sought in such encounters, its arrival produced mixed emotions—triumph but also guilt and fear, as now the person inflicting the wound was made vulnerable to the family members of the hurt or killed person and the state in the form of the police and the courts. Shohidul described how at this point Thandu and his wounded men were carried to a house where doctors had been called. They went to work on the cuts and gashes; unable to staunch the flow of blood, they rushed the men to the local hospital, where Thandu and a few others died shortly afterward.

The preceding story was told to me as an instance of how the men of these parts bravely faced down the usurpers of their land. Yet such pictures of valiant men taking a last stand on their land blurred when one got closer. On a visit to Tangail, the district alongside Sirajganj where people often migrate when their land erodes, I stopped by the house of a local politician, who was originally from Sirajganj and whose father and brothers were closely associated with Nannu, Thandu's nemesis, and fought on his side during the kayija. Unlike most house visits during which I was either greeted with warm welcome or at least with curiosity over what development funds I came bearing, these family members seemed to anticipate my curiosity about the fight and were very circumspect in what they said.

They told me their version of the fight in which the story started at a different point than in the account I heard earlier. It began with the murder of their father, also a politician, in the dead of night in the years following the independence of Bangladesh in 1971, twelve years before the kayija. He *was murdered because he knew the ownership of the land all too well*, said the eldest son. That is, as a politician the father most likely knew about land that had been abandoned by Hindus leaving for India.[13] When this murder took place, he had been working with the government to secure the property rights of those without any land to their name. Although the family did not implicate anyone directly, I was given to understand that Thandu was the prime suspect in the father's murder as he sought the land for himself.

The conversation veered off to that period in Bangladesh's history when those newly returned from fighting the war of independence against Pakistan roamed aimlessly with nothing to moor them to their old lives. Thandu and Nannu were such newly returned war heroes who now toted guns. *Were they friends?* I asked the politician's family. They looked momentarily nonplussed, exchanging looks among themselves before settling on saying, *Well, they seemed to come back from the war as close friends.* So how, then, did they get to the point of fighting each other? And whose land were they fighting over? I thought better of asking these questions and made my exit.

I continued my efforts to understand the contours of the land battle by going to speak to Thandu's nephew, Ansef, who had fought alongside him and was one of those severely wounded at the same time as Thandu. Because he too lived in Tangail, we drove away from the suspicious family of the politician in an attempt to reach Thandu's nephew's house without their knowledge. Instead, we drove around lost in lanes and by-lanes before realizing that the only way to go to the nephew's house was past that of the previous family's. In other words, Thandu's family had to go past their enemy's house, perhaps on a daily basis, in a part of Bangladesh to which neither of them belonged and where neither was compelled to live so close to the other.

Ansef's account was almost as dramatic as that of Shohidul and as circumspect as that of Nannu's political associates. The two, Thandu and Nannu, had been fast friends since the war, in which a motley group of Bengali youths had come together to hurl themselves against the Pakistani killing machine in the 1971 war of independence. When Thandu returned to his village, he was bequeathed the tattered remains of an old claim on land that a group of elders from his village, including his father, had acquired by means of a *potton*, or permission to till, from the zamindar of the day. Although the zamindari system ended in 1956, the group of men retained their claim on this land by means of another potton, this time from the state that had come to possess all of the zamindari land. Or perhaps their claim obtained from the zamindar was superseded by another party who dealt directly with the state. The nature of this potton never became clear to me.[14]

Although Ansef did not say so at this time, we found out later from one of the men who benefited from this fight that the land in question measured 500 bighas (309 acres), of which 110 bighas (68 acres) lacked a dag or plot number on a khatiyan, or record of registry. In other words, these 110 bighas were not legally owned or claimed by anyone and may have accreted only after the surveys of the early 1920s. *Oyi jomindar shob bhejal bajayechey* (That zamindar created all the trouble), said our informant. The zamindar who owned the original plot of land may have simply laid claim on the accreted land without giving due notice to the state, and in granting permission to a group of villagers to till it, he may have passed on the burden of legality to them.

Whether the state was aware of the ambiguous status of these 110 bighas when it came to possess the larger expanse of land in the late 1950s was not clear. Nor could the state really claim the land as one khatiyan, the status designating it as khas, without a survey and settlement of it, and the survey process was notoriously slow. For instance, the land of the village where I resided remained unsurveyed although it had been up since 1988 and had acquired the informal status of qayem. The state may have also passed on the burden of legality to the villagers to whom it leased the land. In time, as land came to be coveted, particularly in the period following 1971, which saw the emigration of many Hindus from the area and the rise of men such as Nannu's political associate who knew the ins and outs of disputed land and was likely murdered for such knowledge, the mounting legal cases and attempted grabs on the land meant that Thandu had inherited a conflict and the promise of a kayija. One might think of the land as having gone khas in the sense of having gone rogue, a somewhat different meaning than in the sense of having gone to the state.

Proliferating Relatives

Just as in the fight in Phulhara, there was a privileging of secondary relatives within the mother of all fights. During my visit with Ansef, Thandu's nephew, Ansef's sister, now an elderly woman, repeatedly slapped her forehead and cried for her *shonar mama*, beloved or literally golden uncle, her mother's brother. She cried for her own loss of a beloved uncle but also cried for his loss, as he had died of heartbreak over his betrayal by his friend Nannu. A different kind of brotherhood, produced in fighting the Pakistani army, also prevailed in this telling.

Gradually a disloyalty toward his mama, or mother's brother, entered Ansef's narrative. Ansef said, *Thandu was determined to win the kayija at any cost. He had hired men from Atapara, Kakuriya, and Makurkul.* These were villages in nearby Tangail district known for their men for hire. Apparently, their short, stocky, powerful bodies were a sharp and intimidating contrast to the thin, wiry bodies of farmers. Nannu had only a few young men, and his political associate, the murdered man, had his *gusti*, or clan, but Thandu's army outnumbered them. *They had come supernaturally strengthened by pora pani*, Ansef said, indicating water over which a powerful kobiraj, or medicine man, had said verses from either the Quran or one of the many books of magical spells on sale in every town. Yet Ansef thought that despite Nannu's unfair use of *kala jadu*, or black magic, it was God's will that Thandu should die. Why? I asked, puzzled that Thandu's nephew would proclaim the necessity of his uncle's death. *If he had not died, then his success would have ensured the entry of musclemen into the area, as land had been promised to them for their support, and that would have been wrong*, he replied. While the land lacked a clear lineage of ownership, this being the condition of possibility for it to go khas, it was not entirely a tabula rasa. It belonged to this village and those who hailed from this village, and to be owned by someone outside the area would entail its metaphoric dissolution or submergence. Here brothers through friendship and belonging to the same village, what is sometimes referred to as *gramer attio*, or relations through belonging to the same village, took precedence over being uncle and nephew.

But this understanding of kinship relations could not stay bloodshed or the death of men, which brought on the full presence of the state in the shape of police officers, criminal charges, and shalish. The upshot of the process was a short jail sentence for some of Nannu's men and a fairly even split of the land between the two groups. Then the land went underwater in 1983, a short year later. The entire village of Doshayiga, its surrounding environs, and the villages adjacent to its two sides, including Shohidul's village, broke off chunk by chunk. The two groups were left holding a resolution to a land that was no more.

Ansef was granted Thandu's share of the land as Thandu's second wife, Ayesha, from Faridpur, who had fought alongside him, returned to her father's village in the more prosperous southern part of Bangladesh. The children of his first wife wanted no part of it. Thandu's younger nephew, Ansef's younger brother, also wanted no part of this land, so he educated himself and moved away to a nearby district to teach religious studies. But before his brother moved away, Ansef married him off to a girl from the family of Nannu's political associate, the murdered man's granddaughter. I asked why he bothered with such a marriage given that the animosity had putatively stopped with Thandu's death and the distribution of land among the various groups. Ansef replied that it was to cement the relations between the families such that should land come up (which he was counting on because, living and fishing with his wife and sons alongside the Jamuna, he was barely making ends meet), such an alliance would help offset further animosities over the land. Thus, he not only expected the land to return but fully expected the conflicts to continue; his brother's marriage into his opponents' family made allies of them, calming their desire for his land or perhaps making it possible that they would help him to face down new opponents. This alliance explained why we had to go past the house of the murdered man and that of his sons in order to speak to his enemy's nephew.[15]

In the Realm of Paper

It may seem as if it was largely through physical fights over land that relations were worked out and that collateral relations in particular—such as those between sisters, cousins, uncles and nephews, or brothers produced of friendship or belonging to a village—were privileged and extended above and beyond the usual powerful kinship figures of fathers, brothers, paternal uncles and cousins, and parents-in-law.[16] In actuality, even the conflicts pursued through the quasi-legal realm of buying and selling, and fighting court cases over land were used to secure relations with those at the margin of normative kinship orders. This was a chaura twist to the Bengali order of kinship conditioned by chauras' negotiated reality and mode of existence. In this section, I follow the ownership records and legal cases pursued by Moinul bhai to draw out his articulation of a mode of care that was patrilineal and intergenerational, of fathers tending to sons and paternal uncles tending to nephews, but within which collateral relations, such as wives, sisters- and brothers-in-law, and nieces, also came to be imbricated in important ways.

Moinul, a wiry man in his early sixties, sporting a beard for no clear religious reason, was a quixotic figure in the village of Dokhin Teguri. He had some standing, evidenced by the fact that he headed one of the three shomajs in the village, groups to which each village member had to belong. As one villager told me, *Shomaj er bairey jibon nai* (There is no life outside of shomaj). To be the head of a shomaj connoted social standing among the villagers although schism was an intrinsic aspect of this formation. Indeed, the two other shomajs in the village had broken off from Moinul's own, suggesting some discontent with him. Some of that discontent was intrinsic to his profession, which was of an unofficial land surveyor, as people most often turned to him to correlate their land records to the realities on the ground. He had undoubtedly upset some people in the process. But some of the dissatisfaction had to do with the fact that he had progressively, in each version of the village, done better for himself, buying land, going into rice cultivation, making his household larger, buying cows, and so on. He had also managed to secure a strong relationship with MMS, my host NGO, by helping it procure the land on which its offices were built.

Not everyone thought well of Moinul because he had done well for himself. Another big man in the village, Tocman, who identified himself as a *grihosto*, or householder, sneered that Moinul and his family had been *kamars*, or blacksmiths, when the two were young in a version of the village from the 1960s, without any land of their own. This implied, first, that Moinul had defrauded landowners through his deep acquaintance with land-related matters gained as an apprentice to a government surveyor during his teenage years. A second implication was that Moinul had no right to any land, acquired either legally or illegally, or to being a grihosto, given his humble origins.

While I was never able to see the full extent of the records Moinul maintained or the parcels of land that he laid claim to by law or adverse possession, what he showed me was quite revealing of a strategy of buying and selling to maintain a line through family relations. On a bright October morning, in the Bengali month of Kartik, already blazing hot by 9:00 a.m., we arrived at Moinul's homestead to begin to look at his documents. He spread out a gunnysack under a shady tree. The excitement in the air had less to do with our presence than usual. Women, children, and passersby were more taken by the rare sight of documents, actual artifacts of writing, in an area where most people were illiterate but where documents commanded a huge presence because of the work they performed in placing people and fields on unmarked landscapes.

When we began to sift through the documents, stained and dog-eared, it was as if we were looking at a meaningless jumble of papers. Even Moinul could hardly keep the papers separate, showing us first one and then another and asking us to read out a few lines from each to help him figure out which properties the papers were referring to. This suggested not so much his modest level of literacy, which was not unexpected, given that he had only been schooled to the fifth class, but that of the scribes who wrote in a scrawl that was unreadable even at the time the deeds were first prepared, some dating back to the British period. For Moinul's purposes it was enough to know the dag or plot numbers, the khatiyan, or records in which these numbers were recorded, the names of the buyer and seller, the amounts of money expended, and the dates that provided the chronology of sales. This information was spread out over the pages, sometimes in the margins but always at the same spots so Moinul's reading ability extended to scanning the pages to elicit this information. He suggested that we seek out another elder of the village if we wanted to read each deed more closely as that person had more mastery over the genre. In a society of few lettered men and women, the work of reading had been specialized, with each person an expert on reading the same documents and books for different kinds of information.[17] Moinul's real talent lay in reading how maps were to be transposed upon material land. In his measurement bag, he carried a tattered geometry set with a small ruler and a protractor, a measuring tape and a twenty-link chain, which helped him to navigate three to four measurement systems, the metric (hectares), the mile (acres), and the bigha-katha (decimal, obsolete everywhere except in land measurements).

As we returned to Moinul's house day in and day out, the papers, in which I found thirteen full records, finally fell into a temporal sequence. The oldest pieces of property had no deeds associated with them. Instead, Moinul maintained what looked like a sheet out of the original British khatiyan, which provided names of the landowners and their plot numbers. To help locate the plots on the ground, these slips of paper also provided the names and plot numbers of those in adjoining areas. The earliest such sheets dated to 1946 when his great-grandfather bought the first piece of property to ever belong to his family, perhaps when he could first afford to do so and not surprisingly at the twilight of the zamindari system, which was effectively over by 1950.

The deeds from the Pakistan and Bangladesh periods were on stamp paper, beginning with 3 rupees in the early years and ending with BDT650 in the most recent one. The turbulent history of the region floated into view through these deeds as the missing stamp papers indexed land bought during the time of the *gondogol*, or the troubles, referring to the 1971 war of independence during which the Pakistani state suspended all services. However, this was the time of the greatest urgency in the selling of lands as people sought to secure some form of cash flow with the complete suspension of ordinary life under the Pakistan army's systematic assault on the Bengali countryside. It was during this time that one saw the start of sales of deeds of land under water, attractive also as a symbolic realm untouchable by the actual flames that devastated the fields above ground. Among these deeds was one that stated that Moinul's uncle sold submerged land to his wife for a paltry sum, but as Moinul explained, it was to compensate her for selling her jewelry to support the wider family. The same uncle also strong-armed his sister's husband—that is, his brother-in-law—to buy several such useless pieces of property around the same time.

While the idea was to keep the property within the family, the dissolution built into kinship relations would mean that Moinul was never able to recover all these lands when he came of age and was financially able to buy them back. Incidentally, Tocman, the farmer who despised Moinul for being an upstart, held some of the deeds to Moinul's family lands that the family had been forced to sell in desperate times. Moinul told me that Tocman planned to never return them despite Moinul pleading with him and offering him more than their market value.

The more recent deeds, the last dating to 2003, showed that the Bangladeshi state had standardized the writing of these deeds as most of the papers came with pretyped text into which one had only to insert the necessary information. This standardization showed how the infamous Vested Properties Act of 1974, amended in 2001, still exerted a spectral influence on property transfers as these deeds had to show proof that the lands were not abandoned property, which was always imminently khas.

Ultimately, however, these records spoke to their greater importance in preserving patrimony, having it provide security of sorts during times of political and personal crisis, above the actual usefulness of the land for cultivation purposes.[18] This is because the documents that Moinul kept so close to himself gave him rights over patches of property that were no more than fractions of acres (the largest starting off as 66 decimals or .6 acres that quickly further fractionated over the generations) and were far-flung. As he confessed, he could not be bothered to cultivate this land. The existence of these dispersed pockets of land and his tenacious hold over their records spoke, rather, to a series of transactions across the generations that tracked a precious, even fragile, movement of guardianship and care, a movement that he held very dear and desired to pass on. The following, in brief, is that line of movement as I put it down in my field notes:

> Darog Ali (Moinul's great-grandfather) died. His eldest son Loku Sheikh took over care of Soker Sheikh and Amir Jan bibi (his siblings). Before Loku Sheikh died (Amir Jan dead already) he told Soker, his brother, to take care of Moyijuddin Sheikh (Loku's son and Moinul's father), along with Shontesh, Soker's own son. So Moyijuddin and Shontesh, first cousins, lived together as brothers. Soker died leaving behind Shontesh and Shukurun (his son and daughter) charging Moyijuddin (his nephew) with their care. At Moyijuddin's death, he put Shontesh (his cousin) in charge, asking him to take care of Ainul, Moinul, and Jobida (Moyijuddin's two sons and one daughter). Ainul died before Shontesh could pass on care of his own family to him, so when Shontesh died in 1993 he passed on the care of Alomgir, his son and Moinul's first cousin, to Moinul, and that care continues to this day and, Moinul hopes, into the future such that Alomgir will tend to Moinul's family once he passes on, but he doesn't seem very optimistic. *Alomgir is not a good man*, I learn from Shohidul.

The line, from Darog Ali, the patriarch of the family, to his children Loku Sheikh, Soker Sheikh, and Amir Jan bibi, then on to their children Moyijuddin Sheikh, Shontesh Sheikh, and Shukurun, further to their children Ainul, Moinul, and Jobida traversed a familiar patrilineal line. But it was also one of intergenerational care recorded in land transactions. It was often the case that when I was walking the chars with people, they would for no apparent reason launch into very sad episodes from their lives involving the love of a lost elder or the loss of a child, as if this landscape, effectively new in contour and configuration in each iteration, was still awash in memories and pools of affect, as if further fighting their potential characterization as khas. In a similar manner, while sifting through these documents, Moinul would often launch into moving stories of the love his uncle, Shontesh, bore for him. He recounted, for instance, how his uncle would place little Moinul on his shoulder and walk miles with him to a relative's house to get him food if his own wife begrudged giving food to the orphan. From these records interlaced with memories, I came to appreciate the substance of the patriarch that imbued these lands and the documents associated with them, making them symbolically important for Moinul to fight for. But I also took note of the importance of wives selling their jewelry to support a husband's wider family, and sisters- and brothers-in-law buying worthless land deeds, in making this line of association workable, particularly in times of crisis such as the 1971 war of independence.

In the Land of Sisters

My specific focus on kayijas and legal battles has been to show how different parts of the universe of Bengali kinship were animated within the context of chars beyond what the normative order would suggest, most notably that of collateral relations between sisters, matrilateral male cousins, uncles and nephews, fictive brothers, wives, sisters- and brothers-in-law. Where were women in this picture, as they could also potentially enter into kinship relations through the rights given to them by Islamic inheritance law and through their own capacities for buying and selling land? How did land mediate their relations?[19]

The literature on women and property in Bangladesh informs us that women are mostly deprived of their inheritance from their fathers (M. M. Rahman and van Schendel 1997). As noted earlier, daughters are entitled to half of a son's inheritance according to the Muslim Personal Law Application Act (1937) and the Muslim Family Laws Ordinance (1961), but more often than not Bangladeshi Muslim women disclaimed their rights and often did so voluntarily. The most common explanation given for this was that women sought to retain a relationship with their natal families so as to be able to depend on their fathers or brothers for protection in the instance of marital discord or early widowhood.

The exclusive focus on women's loss of inheritance denies the emotional and transactional labor that women often performed to support their natal families throughout their lives. In other words, in giving up any land that they stood to inherit or claims on any other inheritance, chaura women could be seen as not only strategizing for an uncertain future but also ensuring strong ties to and claims on the natal home for themselves and their children. To suggest this is to render them as much more self-interested and calculating than the thickness of their relationality would allow. Rather, their renunciation must be considered as part of a family of gestures, such as fighting kayijas or bequeathing guardianship, by which they too attempted to inflect certain relations as more important than others within the universe of Bengali kinship. Consider a few of the female figures in the events I have recounted in this chapter. In the almost-kayija in Phulhara we were alerted to the fact that this might have been a fight between two male cousins by the figures of their sisters who cried for the pain inflicted on their brothers rather than any land lost. Theirs was a performance of a mutuality of being. Of Thandu and Nannu's kayija from the 1980s, it was Thandu's niece who cried for her uncle's loss and estrangement from his friend, yet she supported her brother Ansef's exclusive bid for the land once it reemerged. A scrutiny of Moinul bhai's land records and his narratives showed how Shontesh's wife sold her jewelry to support her husband's family, and Shukurun, Shontesh's sister, and her husband bought deeds from the family when the land was in the river with the promise to sell it back to the family when the land reemerged, which they subsequently did.

In the forty-five surveys I conducted in households across the three villages I had selected in Sirajganj, about half the women heads of household said that their fathers had no property and the women had no expectations of inheriting anything upon their fathers' deaths. However, among those who did stand to inherit, there were very interesting differences in how women disclaimed any rights of inheritance. A significant number said that no one in their gusti (indicating their husband's clan) brought their *moyuri*, a term indicating the land that the women could rightfully inherit from their fathers. The women or the men speaking on their behalf always articulated this statement in a very prideful manner, suggesting a close identification with their husband's lineage or a voice overwritten by the voice of the lineage.

Another equally significant number said they could not lay claim to their inheritance because they didn't want to hurt their brothers. They made statements such as I don't want to snatch away my brothers' joy or I don't want to give any hurt to his heart, implying a personal closeness to their brothers. In a few cases the women complained that their brothers had forcibly possessed the women's lands, again providing an intimate picture of the kind of violence being done to them, with comments such as Amar warish boyiya khaytechey (They are engorging themselves on my inheritance). Some women said they would not make any claims on their property while their parents were alive, while the lands were underwater, or while they themselves were alive, leaving it to be understood that they might make claims later or else their children would do so.

These varied responses suggest that women's renunciation of their rights to paternal inheritance cannot be read back to a single motivation of seeking to secure their links to their natal home as a form of insurance. Rather, the heterogeneity in their tonalities suggests that for some women, their relationship to their husbands' lineages took priority over any ongoing relations with their natal home. For others, it was the love for their brothers that motivated their renunciation, whereas for some, their brothers' greed and lack of love for them accounted for their loss of inheritance. And for yet others, the fact that their parents were still alive made them disinterested in their inheritance. Just as in the case of kayijas and intergenerational bequeathing, control over land or the renunciation of it provided opportunities for women to privilege specific relations over others, to make certain kin matter more than others through the mediation of land.[20]

Women were not entirely spared the violence of kayijas in disputes over land. On at least two occasions in my field study, women were involved in very violent disputes with their natal homes for rights over their inheritance. In one, the woman was obviously not as interested as her husband in her inheritance, but she had to bow to his interests, on account of his having a "hot head." In a second case, Nurjahan, a woman living in Dokhin Teguri, came to show me the top of her head, which had been bashed in by her half-brother from her father's side by a *haturi*, a stone grinder used in the kitchen. Nurjahan had made a successful bid for her portion of an inheritance from her half-brothers and had secured not only her own lands but also those of her sisters. However, since her sisters did not live in the village and had managed to lease or sell their lands, she became the sole object of her half-brothers' anger. They terrorized her, driving their animals onto her cultivated lands, laughing at her behind her back, and stalking her at night until she felt that her life was a living hell. Nurjahan could not even properly mourn her eldest son, whom she lost to a dog bite, because of the myriad ways in which these half-brothers tormented her instead of offering love or sympathy. *How could they not feel sad?* she shouted. Her son was also their nephew, who had helped them in countless ways in their fields. She suspected that they were mentally ill.

After one half-brother brought his haturi down on her head, she was able to bring a case against all of them. Unfortunately, one of her key witnesses, presumably bought off by her half-brothers, provided weak testimony. He reported that she had been hit by a mere stick instead of a haturi, causing her case to be thrown out of court on the grounds of it being a false charge, but Nurjahan still gloated that she had managed to keep her half-brothers in prison for up to a month and had forced them to drag themselves to court for another month. Meanwhile, as she related to me in tears, her doctors had told her she ought to enjoy what life she had left, which amounted to their saying they could do no more for her and that she had to live with the threat of her imminent demise. When I told my friend Kohinoor about Nurjahan's condition, she scoffed, saying, *The poor don't die easily*. Indeed, every time I returned to Dokhin Teguri, I would seek out Nurjahan, dreading news of her death, and I found her very much alive, now with a son having gotten married, now with a granddaughter, now owning a sick cow, now with a whole herd of cows to her name.

Although Nurjahan's conflict with her half-brothers did not merit the designation of kayija, which seemed to be saved for major conflicts between men out in the open, she showed me a list that she and her husband maintained of the many confrontations, shalishes, and court cases she had been involved in with her half-brothers to date. In a perverse kind of way her list suggested this was the part of her kinship universe that she had inadvertently ended up privileging over others. This choice had been thrust upon her and wasn't entirely of her own making, but it was how she had chosen to live.

Conclusion

In this chapter, I have attempted to show how a history of land laws and the market for lost lands produced the conditions of possibility for owning land even in its absence and for perpetuating and circulating land even with unclear ownership, as well as created the conditions for a conflictive sociality in the chars. The chauras were clearly fighting over a rare resource in the midst of want and state negligence. At the same time, the fights were not entirely controlled by the puppet masters of old, be they the state, absentee landlords, or locally based established landowners. Nor were the fights indiscriminate. The actions taken were understood to be within the range of acceptable violence, from fighting with sticks and bricks to legal cases.

In the process of fighting over land, the chauras animated new parts of Bengali kinship, determining the mutuality of being—that is, who was implicated in another's existence and how. While at first glance it would appear that these fights were instigated by patrilineal patrimony, or claims on land through inheritance from one's father, and therefore animated those who claimed a common descendant and residence, such as brothers, upon closer scrutiny it would seem that fights also involved many collateral relatives, those less important to the normative kinship structure. This insight led me to explore the expansion of such kinship through one's mother's brothers and sisters, matrilateral cousins, and one's sisters and their husbands, suggesting that this particular line of kinship relations was most productively reanimated and strengthened through land-related fights in the chars. In turn, these collateral kin came to bear the burden of keeping guard on land in its absence and of handing it over should it emerge from the river.

I have also sought to slow our judgment so that we could see that when chauras participated in the legal fiction of absent land as property, this was not simply the product of the contingencies of history, complicity with the state, or kinship driven but, rather, indicated a relationship with the land itself to extend its presence even in its absence. It took mental labor to maintain land just as surely as it did physical labor. I take seriously that what the char dwellers were giving expression to was char land filling space not by its existence alone but by its extension through chauras' dreams, speech, and actions. In other words, matter extended itself through mind.[21]

The lands expressed tendencies, and chauras apprehended and actualized those tendencies. As predominantly Muslim, they subscribed to the idea that God was the creator of the world and that everything submitted to divine will, even nature and humankind (Thorp 1978, 1982). At the same time, as a normative religion, Islam also prescribed an "oughtness" to every domain of life.[22] This oughtness derived not only from revealed texts, the example of the prophet Muhammad, and religious guidance but also from signs in the surrounding world, with physical nature's regularity taken as proof of the perfection of creation. So although humans were considered to be ignorant of all the mysteries of God's creation, nonetheless Muslims had been endowed with aql, or reason, and were bid to study, observe, and learn to elicit normativity, teleology, and intelligibility from the natural world.[23] Through aql they posited that matter was drawn to like matter, and char lands sought extension through chaura existence and activity.

Two young men of Phulhara laughingly demonstrate how fighting is done with lathis. Photo by author.

Moinul bhai holds up a land deed on British India stamp paper in his possession. Photo by author.

History and Morality between Floods and Erosion

The Nation-State and the Everyday

THANDU'S BODY HAD BEEN IN ITS EARTHEN GRAVE less than six months when Nannu and his associates were released from prison after they promised to work out an understanding with Thandu's family over the land in Doshayiga to which both sides laid claim. But no sooner had the patchwork of claims been sorted in a shalish than Doshayiga broke, leaving everyone with only an oral agreement. Thandu's second wife, Ayesha, left for Faridpur in the south of Bangladesh; Thandu's children by his first wife turned their backs on the event; and Thandu's nephew Ansef was left to live by the riverside, fishing and planning for the eventuality of the return of the land, having given his brother in marriage to Nannu's political associate's family.

What happened to the other villages and houses that slipped into the water when Doshayiga broke? I spent a long time during my fieldwork tracking the movements of people in the wider area in the event of riverbank or land erosion (bhangon) as opposed to the other major weather event that often left them landless—namely, floods (bonna). While char dwellers could recount where they were and what they did in every instance of a flood that displaced them from their land, they were less certain of their movements in the instance of erosion. One ready explanation for this was that floods occurred less often, with a periodicity of five, ten, or up to twenty years, whereas land erosion was ongoing, with unstable soil shifting constantly, succumbing to the pressures of the river waters and of human settlement and cultivation by breaking off in bits and pieces or large chunks. Floods were total events that swept aside everyone in their way, whereas erosion affected only a few people at a time unless or until entire villages collapsed.[1]

The narratives of floods were also notably different from those of erosion in that people wove their individual experiences of floods together with national events, often events such as the collapse of a military regime, which were spurred by the floods themselves due to the inadequacy of the state emergency response.[2] In contrast, the narratives of erosion were halting and uncertain, with large segments of time unaccounted for. An analysis of my fieldwork data relating to floods juxtaposed with erosion leads me to make several claims. Floods, however small or large, were national events. Since its foundation in 1971, Bangladesh had crafted itself as a nation-state capable of dealing with two kinds of crisis, notably floods and famine. When char dwellers recalled their lives by referencing major floods, they showed themselves to be historical subjects, evincing awareness that floods punctuated the history of the nation-state, interweaving their lives with this historical timeline. But when chauras failed to recollect their activities during the time of erosion, they showed themselves to be subjects obscure to themselves. A persistent dimension of the time of erosion that I observed was the spirit of cooperation by which people helped each other break down their homes for transport elsewhere, the acknowledgment of the need for resettling those without homes, and an ethos of reciprocity and hospitality. Yet very little of this dimension was recalled in the aftermath of erosion; in fact, people groused that they had been badly treated by both intimates and strangers.

One understanding of this forgetting is that erosion was of the order of the everyday lined with skepticism (Das 2006). It was not only an occasion of the abrupt loss of one's familiar world but also a period of disappointment with one's intimate others despite what they might do. Erosion accentuated separateness. An additional reading of this self-obscurity is that people became a different kind of subject during erosion, not just obscure to themselves but acted upon by the landscape to which they gave themselves over as if they were sediment acted on by the river. Entrainment, as this process is called within geomorphology, involves the action of water on sediment, producing ripples on the river floor, or the suspension, channelization, and deposition of sediment along the length and breadth of the river that gives the river channel its specific morphology (Church 2006). Lest it seem that this is merely a physical process and humans are not so easily dissolved to the status of sediment, it is helpful to consider the observation by Garry Peterson (2002) that there is a tendency for landscapes and ecosystems to reproduce particular patterns after a process of disturbance. This pattern is far from a state of equilibrium and suggests only that systems carry strong ecological memories of preexisting states. In other words, if we think of the chauras and the river as being in a particular relationship within the wider ecosystem of which they are both a part, then a disturbance in the river jogs the system's memory and prompts char dwellers' reproduction of past actions and patterns.[3] One might provocatively say that those in the throes of erosion were not self-conscious, much less historical, subjects. Rather, chauras' obscurity to themselves indicated their state of receptivity and thereby their transmission of the river's activity.

These observations find echoes in Schelling's *System of Transcendental Idealism*, in which he writes, "In all producing, even of the most ordinary and commonplace sort, an unconscious activity operates along with the conscious one" ([1800] 1993, 228). He explains this unconscious activity of nature as a "dark unknown force" within us: "This unchanging identity, which can never attain to consciousness, and merely radiates back from the product, is for the producer precisely what destiny is for the agent, namely a dark unknown force which supplies the element of completeness or objectivity to the piecework of freedom" (222). What is noteworthy in what Schelling says is that nature, far from being the hidden hand of human acts, as the unconscious is a necessary "something more" that gives human acts their sense of completeness, even appropriateness.

In this chapter, I begin with a scene of erosion in the chars, using it to unpack the differences between representations of floods and events of erosion. I explore how floods (and famines) intercalate chaura lives with the history of Bangladesh. I next turn to topography, specifically gradient, surface, and subsurface, that interrelates floods and river erosion, to continue to draw attention to the dynamics of the moving land started in the previous chapter. The topographic underpinning of chaura experiences of erosion is next drawn out. The understanding of entrainment helps me illuminate the aspects of chaura self-opacity as receptivity toward the landscape and shows how the many acts of sustenance, care, reciprocity, generosity, and hospitality that derived from kinship assemblages and everyday relations also had an impersonal dimension. Such acts indicated the morality extant during erosion as the awakened ecological memory of the wider ecosystem of which the river-char landscape was a part, or the index of nature as the unconscious of the chauras.

A Scene of Erosion

Nojrul Mistry (Nojrul the Carpenter) crossed my path one evening in the Nauhata Bazaar where men and the solitary anthropologist gathered for tea and news, and children and dogs searched for treats. He was back in the village after his stint working abroad (any place away from this place being equivalent to going abroad) because there was plenty of work around here, he reported. It being the rainy season (borsha, between June and August 2012), the char was prone to natural disasters. While intense rains interspersed with long dry spells increasingly marked the monsoons,[4] fears mostly co-alesced around floods (bonna) that would be more or less than anticipated.[5] This mixed desire for and fear of rising waters was compounded by the sense that there was an alchemy of water; it had to be just so to make for a successful agricultural season, although accounts varied on what that should be.

But there was no work for a carpenter during the rainy season because of the floods. Most chauras characterize the rainy/flood season as one involving mainly sitting around. What work there was resulted from the opening up of cracks on the ground within households or the sliding of banks to the river's edge, indicating that time was short before villages would slide or drop into the running waters. Nojrul was here to help dismantle buildings, although the word seemed disproportionate to the mobile, worm-eaten, half-rotted structures that housed people, their animals, and their supplies and that felt barely worth propping up again and again. His know-how as much as his tools could quicken the work, so Nojrul's services were in high demand. He prioritized based on the urgency of the demands and the familial claims on him. And he maintained a rough distinction between those to whom he was related by blood, who were his shoreek, to whom he owed immediate assistance, and those to whom he was related by marriage, kutum, or fictive kinship, attio, from whom he could expect remuneration.

Early the next morning, Nojrul was heading to the southeastern part of the island where the villages of Koayrat, Baliakandi, and Bantior were slipping into the waters. He was going to help a family member, although it was understood he would be paid for his work. At the time of my field-work between 2011 and 2017, the typical day's wage for a laborer was BDT300 to BDT500 (USD5), but he expected up to BDT1,000 (USD12). He offered to have me come observe the preparation and process of bhangon, a catchall term that captures the physical process of erosion, the dismantling of one's household, and sometimes one's own physical and mental breakdown.

Although Shohidul and I were up early the next morning, Nojrul Mistry had left before us. We understood why. He may have caught an unexpected ride on someone's boat that would take him directly from his house to another farther along, as water had submerged most of the pathways, making them difficult to traverse by foot, but making the journey much easier by boat. This water that covered the low-lying areas of Dokhin Teguri in July was not described as floods. It was too variable for that, with its level changing from day to day. It was explained to me that either rainwater had gotten caught in the dips in the ground or else river water slowly rose and seeped up through the soil (fluffed by snowmelt to the north in Nepal?). It was only when the two occurred together that there would be floods, which could only be gauged by a precipitous rise in water level, more than two *hath* (hand) by chaura estimation. A hath, taken to be eighteen inches, was an obsolete form of measurement, one among a plurality of systems that coexisted in the char, suggesting measurement as a palimpsest of history.

We hitched a ride on the boat carrying workers of the NGO where I was housed to one of their drop-off points from which they could go to collect loans from women across the island. Making our way from the drop-off to where Nojrul was likely situated meant walking some distance and grabbing further rides on boats until, three boat rides later, we were standing on the edge of the island that was now expected to slide into the water. Along the way we stopped to admire many trees and shrubs that we didn't usually see in the village where I stayed because the soil here was older, richer with humus, and more fertile. It was a shame to see such good soil go underground.

The edge itself was an area of low-lying verdant fields with the household where we were headed beyond it. Although raised on a mud mound above the low fields and surrounded by tall trees and bushes, the household stood facing the river waters. This arrangement felt instinctively wrong as households are usually at a distance from the river's edge, with crop fields and open grassy lands between them. It seemed as if one part of the picture had been lopped off. On closer perusal, it looked that way, as the river's side of the household was a jagged edge with clumps of earth and exposed tree roots dangling perilously. The fact that the very arrangement of the landscape felt skewed led me to reflect on the aesthetics of rural life in Bangladesh that naturalizes the agricultural over the deltaic.

The household was not just one but several households that had once belonged to a patriarch and his three sons who had striven to live together on the five bighas (approximately three acres) of the patriarch's ancestral land that had come up some ten years ago. Because the quality of the land was not particularly good, they erected their homes on it, taking adjoining land on yearly rent (khajna) for cultivation. Other family members had joined them on their land, bringing the total household number to eleven. This play of numbers was very important (four heads of families, five bighas of land, ten years, seven additional families) as it suggested the constant calculus that made up the mode of existence on chars in which both the qualities of the land and the claims of family were highly variable and interdependent. But now the families were dispersing, with each going in a different direction to wherever they could find refuge (asroy). Leasing land on the mainland was out of the question because of its exorbitant rates, with high deposits up front in the form of kot (BDT30,000 to BDT40,000 per bigha or sixty-two decimals or two-thirds acre). Although yearly leases for minimal amounts (anywhere between BDT500 and BDT4,000 per year for twenty-four decimals or one-fourth acre) were the norm within the chars,[6] increasingly landowners in the char were beginning to ask for deposits up front or even refusing to give land, which went against every code in the chars.[7] Consequently, there was no possibility of these households making the move together, although it might be possible for them to conjoin after several such moves or perhaps after their own land returned or was returned to them by the river. Both of these possibilities hung in the air, conveyed through common chaura sayings such as Land seeks to return to the river or The river throws up land as a boon to humans.

Several empty plots with some remaining bundles of twigs and branches indicated that a few of the households had already left. Nojrul was busy at work on the roof of one household while several others remained intact as their owners continued to seek places to which to move. One was too burdened by its weaving factory to know where to go. Although a weaving factory, which was simply a shed with several looms, might indicate more skills and an extra source of income for households in the char, it also translated into greater difficulties in moving. One not only had to negotiate more space than most landowners were willing to part with but also had to insert oneself into a local economy of weaving that already might be riven by rivalries and saturated with competition.

One household didn't yet have the wherewithal to move because illness had earlier struck the wife of the head of the household. That had cost them the necessary capital (*shomortho*) to move. Naznin, the wife, stated forlornly, *I am parentless* (*ma-bap chara*), indicating that she had fewer options than others for where she and her husband might go to seek shelter, suggesting the renewed significance of a wife's natal home within the context of the char. Elsewhere in rural Bangladesh she would have returned to her parents' home upon facing abuse or abandonment by her husband, but in the chars she could have returned with her husband with no shame in the event of floods or erosion because it was her parents who had tied her fate to the river, and they were responsible for her. But this orphaned woman did not have the option to return. Nonetheless, she dropped the name of an authoritative figure in Nauhata, the village north of my temporary home, which reassured me that she, as it is said, *had caught the hand of someone* who would likely find her refuge, even if it was only on a new char as yet without human settlement. While the women of the household that was moving were busy making the necessary arrangements, this forlorn woman kept me company and provided commentary as if she too were watching a scene unfold.

Nojrul was busy undoing the corrugated iron roof (*chaal*) of the main house of the household. After he removed the roof with the help of several able-bodied men, he set to work on opening the sidewalls of the house made of thatched bamboo (*dhornas*) and pulling up the pillars, some of which were made of bamboo; other pillars (*khaam*) were made of cement. He found that white ants (*ui poka*) had attacked several of the bamboo pillars and the wooden frames of the windows and door. The female head of the household was immediately shown these damaged pillars and frames so she could decide what to do with them. If she had insisted they be used for reconstruction, they would have landed on the growing stack of materials that were going to be placed on the heads of about six men, some family members and neighbors but mostly day laborers (*kamla*), who would transport them to the motorized boat (*shallow*) that waited by the side of the fields. She decided, however, that the pillars and frames were too far gone and would be best utilized as fuel for cooking (*khori*), so they were broken into pieces and tied together to make up bundles to be transported later. Young boys, some seven to eight of them, were given the task of making up these bundles. When I asked Nojrul about the boys, I was told that they were *moyirati*, workers who had been brought on with the promise of receiving three square meals. When I asked a young boy what had brought him from home to help, he giggled and said *faram*, referring to the fleshy, farm-raised, feed- and hormone-stuffed chickens that had flooded the national and local markets, making local chickens a rarity. The young people increasingly preferred farm animals because their flesh was soft and easy to eat, unlike the tough, stringy meat of local chickens.

Younger women were bunched to one side of the former household, busy with cooking to feed this army of helpers—some fourteen men and boys to break down the household over a period of two days, and later about five or six to set up the household over another two days. Some of the women plucked chickens, some cut vegetables, and some pulled freely on the leaves and branches of the trees surrounding their home to feed the two open stoves blazing in the corner. In the midst of this busyness a well-dressed woman wearing a sari woven with gold threads sat on a wooden chair cradling a newborn infant with eyes made up thickly with kohl. The daughter of the household had recently returned to give birth in her natal home, and her mother-in-law had come to collect her and the baby. While the new mother was busy cooking for the workers, the baby lay happily in her grandmother's arms, taking in what was affectionately described as her first bhangon. Like many chaura women, she would grow up recounting how she experienced her first bhangon while still on her grandmother's lap.

Gathered and placed in heaps in the former courtyard were the usual features of such households. This particular household was more prosperous than others as indicated by the fact that its members had storage containers, hinting that they had something to store. They had big trunks to protect clothes, bedding, and anything else of value from rats and rot; large woven baskets (*dhols*) to secure their dry goods (rice, pulses, spices, vegetables such as onions and garlic, seeds); mud pitchers (*kolshis*); plates and pans bundled in fishing nets; and the ubiquitous wooden bed (*khat*) that itself functioned as a refuge when houses filled up with water during floods. A portable mud stove stood ready for the move to help them with cooking in the new place. Cows and goats stood around their feeding station (*gora*). The animals would be transported at the very end, as they needed to go to a prepared situation, being quick to contract diseases if not properly fed and if made to stand around in mud and filth. The chickens, however many remained after the cooking was done, would be taken away immediately or else they risked being snatched up by foxes or lizards. The pigeons kept by the family took badly to transitions, tending to return to the place they considered their home at the first opportunity. They would be moved but kept secured within a birdhouse (khop) during transport and then forcibly kept within the khop for a few additional days to sufficiently disorient them and allow them to become acclimatized to the new place. An old white-haired man sat on his haunches keeping an eye on the proceedings in front of him. He was possibly the patriarch whose land kept bringing this family to this place but who sadly looked like yet another bundle to be transported.

Despite the likelihood of erosion, there was no sense of panic in the air, and people worked with focused attention. The male householders wanted to get the material onto the boat to begin transport. The day was getting on. It was already noon, and the boat still had to make its first trip. There was no fear of rainfall as the sky was bright and blue, but if the day passed too quickly and they were still transporting their possessions, they were worried lest they be left facing the river. They were not worried about its ravages, which had already unfurled and would unfurl in their own time on the remaining land. They were worried that the river now came right up to their household, giving easy access to robbers and bandits. The fear of bandits surfaced as quickly as the fear of bonna or bhangon in the chars. The stories I heard about how bandits struck when the chauras were at their most vulnerable, robbing them of everything, sometimes even stripping them of their clothes, were enough to chill. On several occasions, the char dwellers said, they had rescued people from the remotest edges of their char who had been robbed and left stranded in the midst of tall, dense catkin grass considered to be the haunt of ghosts.

The women of the household worried about other things. The female head of the household wondered if they would be able to put up a fence around the site of the new home so that the women could be safe from the prying glances of passersby. Although they would be getting their food from a neighboring house, she wanted to take her daughter-in-law over early in the day so that the young woman could get acquainted with her new neighbors, having never met them before. The older woman already knew them, having crossed paths with them during a previous stint of bhangon.

Both men and women worried about the trees. While they had no choice but to cut them down to use for their new homes or to sell as timber, they worried that the trees appeared to be slipping away with the soil and that efforts to secure them with rope and twine might not be sufficient to keep them in place for another day and another effort at wrapping things up.

Meanwhile, the fact that the trees gave refuge to *babuyi pakhi*, or weaver finches, whose elaborately woven nests hung ten or fifteen to a tree like small communities, seemed to pass them by. When I inquired after the finches, the women shrugged and said, *Even birds have to experience bhangon. It is everyone's lot in these parts.*

Naznin, the forlorn woman, and I had joined other observers at the edge of the household. A few had come with a fistful of rice, others with small packets of salt, and some others with a few green chilis. Now that word was out that this household was leaving, people had come to pay off the debts they owed in the form of small requests that pass between neighbors. Or perhaps they had come to provide an opportunity so that debts owed to them might be paid because, as the householder who was moving said to me, It *is the worst thing to leave something due, to leave people speaking ill of you.*

What I was seeing in these small transactions was the disentangling of ten years of inter-stitching. It struck me as somewhat cold that people were moving away without apparent concern for the fates of the intimates they were leaving behind. They seemed more concerned that they not leave behind any debts. Although I knew of other instances in which debts went unpaid, loans defaulted, and promises were broken, the insistence on clearing one's account before leaving a place reminded me strongly of the emphasis placed on doing the same before one's death. The answer that was most often given to the question in my household survey about what people hoped for in their future was that they hoped to die with their heads held high, owing nothing, with no claims on them. This resonance between the householder's words and the quotes from the survey suggested to me how bhangon was experienced as death, as an excision from a place, time, and milieu. The stuff left behind, the worn clothes, the torn rubber slippers, the broken stools, the empty bottles, the cracked plastic buckets, the twisted eyeglass frames, the empty box of condoms, and the still-warm ashes of the stoves, gave a profound sense of a life worn by use and then hastily abandoned.

The Imminence of Floods

This particular household was moving to Dokhin Teguri, my own home away from home. Upon arrival at Dokhin Teguri, their stuff was thrown on a corner of a crop field, which was low lying but next to an upraised path that snaked through the village. When I stopped by later that day, I saw that one of the walls of the house had gone up and a temporary roof had been erected, affording some privacy for the women and protection from the rains. The ubiquitous wooden bed and a few bamboo mats were laid out under the roof. I imagined that bodies would huddle together for many nights until enough soil was dug up to create a mound on which to place the more permanent homes.

I had thought I would be able to go regularly to see how the family adjusted over time, but the very next day a heavy downpour flooded the path, leading the family to disperse throughout the neighborhood to take shelter. It would be a while before this household became established, but when I asked if these further waters signaled floods, I was told I hadn't seen anything yet. As the days slipped into weeks and it was time for me to leave by late July, I still had not witnessed what were officially designated as floods. As I was quickly acquiring the macabre reputation of one who wished for floods, I stopped asking. In retrospect, it seemed as if we were always either in the aftermath of a flood or anticipating one. When I arrived in June 2012, I heard that I had missed the sudden floods in May, which—coming as they did much earlier than expected—had delayed the planting of sesame (teel) and aman rice. While I was there over June and July, the water levels kept rising and falling, leading to a very poor crop of sesame but leaving the aman intact. After I returned to the United States in late July, I received phone calls the next month in which I was told the waters had risen almost to the height of the bazaar and that the aman rice paddy was now destroyed.

There was always a sense of urgency to determine whether or not we were in a flood situation. While the villagers were reluctant to pronounce a flood until they had gauged the effect of waters on crops, the politicians and NGOs wished to declare floods on the basis of the egress and presence of excess water in everyday life. The entire time I was in the field I received phone calls from the head office of MMS with requests for me to verify whether Dokhin Teguri was flooded because only then would the organization be able to write missives to draw the attention and, it was hoped, the funds of foreign nations and donor agencies. While there were national disaster management principles, policies, and bureaus, these were seen as being responsive only to those events that reached the scale of a "national-level disaster." Thus, a distant nation and its apportionment of extraordinary funds could be more reliably counted on in the instance of a local flash flood or a sudden bank-line erosion than one's own state. It was rumored that there was a national relief fund for such occasions, but only ministers had the powers to allocate these moneys. One day in Chauhali, the seat of the subdistrict government that was in the mainland, I watched as bundles of money were prepared by local officials of the Awami League (the ruling political party) to give to the minister Latif Biswas to distribute to flood victims. His visit was like that of a celebrity, surrounded by bodyguards and media persons. I followed this crowd until the minister boarded a large boat and chugged off to bring succor to a few privileged flood victims, perhaps from the rumored relief fund.

Later, when I asked Nojrul Chairman, a senior char-based politician who had previously served as chairman of the union parishad, or council, why erosion did not muster the attention of floods, he pointed out that bhangon never reached the scale of floods in terms of producing widespread suffering or media coverage sufficient to warrant national and international attention. At any rate, people generally anticipated bhangon and moved before they could experience its worst effects. In his explanation, people in the throes of bhangon experienced only harm or damage (khoti) and not suffering (koshto).[8]

This very measured assessment of the difference between floods and erosion by someone who had experienced bhangon no fewer than fifteen times suggested how entrenched a population-based perspective was in politics and everyday life in Bangladesh. Floods mattered because they involved large numbers of people and property. As such they appeared to leach the experience of bhangon of expressivity as victims of erosion kept their eyes focused resolutely on the skies and the water level—that is, on floodwaters— giving short shrift to the earth moving beneath them, robbing them of their fields, houses, and pathways. Yet their splayed feet that gripped the soil with their toes and physical postures poised for flight as they walked or sat on their haunches suggested another vector of orientation running through their bodies that was turned toward the ground. Guided by their bodies, I examine chaura memory of floods to understand how floods came to occupy their attention, before turning to their bodily orientation toward the sudden movements of the ground beneath them.

Floods and the History of the Nation-State

When one asked elderly chauras which floods they remembered readily, the ones that were most immediately mentioned were those of 1988. There had been two large floods since, in 1998 and 2007, not to mention innumerable ones in previous decades, so what was so significant about the floods of 1988? That was the year when the floods of Bangladesh took the international stage. Subsequently, there was a joint declaration to stop the ravages of floods in Bangladesh in a G7 summit held in Paris that year that helped launch perhaps one of the most ambitious projects to study and propose flood control measures for the entire country (Brammer 2004; Adnan 2008). The ensuing Flood Action Plan (FAP), a multidonor initiative implemented through the World Bank, has had many critics whose critiques have been pitched less at the limited results the plan finally produced when it wrapped up in 1993 and more at the full scale of damage it would have inflicted if its engineering imperatives had been faithfully followed. But what ultimately made the FAP a bad flood measure was that it was put forward as a fait accompli to the country by a much-disliked dictator, Muhammad Ershad, who, unbeknownst to him, was in the final year of his dictatorship. He was toppled by a groundswell of protests against him in 1989, and work on the FAP would stop shortly afterward.[9]

For all the problems associated with the FAP, its lasting legacy was the forceful introduction of remote sensing and satellite imagery in the study, analysis, and projection of environmental problems, specifically floods and river erosion. While space technology had been in use in Bangladesh since 1968, it had been the preserve of the government. The FAP initiated the establishment of organizations such as the Center for Geographical Information Services, which enabled such satellite technology and imagery to become more mainstream. Along with such access came the understanding of the extent of char formations within the major rivers and the coastal regions in Bangladesh and the durable populations of people living on them (EGIS 2000; de Wilde 2011). One might say that the chauras came into view in 1988, and not a moment too soon, as some parts of the FAP had aimed to protect the mainland at the expense of chars.

My chaura informants didn't only recall the floods of 1988 because the floods spelled the rise and fall of a megaproject, the demise of a dictator, or the chars' emergence as a feature class within a geographic information system map or a rallying point for anti-FAP activists and NGOs. They recalled them because it was the year that large tracts of land they would call their own emerged from the river. Displaced in the 1950s by the Jamuna's migration westward, they had been living along the riverbanks or on embankments for a long time. Suddenly they were witnesses to the possibility of reconstituting their fathers' villages. In order to do this, they had to take leave of their stable identities as rural farmers, which is what they were when their villages fell into the waters, to become island dwellers by moving into the heart of the river to take up residence on a char.

While the rest of Bangladesh was submerged, facing the beginning of the demise of the country, or so the experts said, the chaura population I worked with in Dokhin Teguri was just coming into its own. Their memories of this total event were bittersweet, both brimming with the intense suffering brought on by living underwater for almost a month, the longest time they had perdured with floodwaters until that point, and with the cautious excitement of new beginnings on land that was once theirs.

The memories of the chauras did not linger on the floods of 1988 for long before people recalled another period of intense suffering brought on by floods. These were the floods during the time of shadhin, the chaura shorthand for referring to the 1971 war of independence by which East Pakistan was liberated from West Pakistan to become Bangladesh. This recursion wasn't produced by similarities in the circumstances of flooding, as the floods of 1971 were minor in comparison to those of 1988, in which nearly 65 percent of the country was underwater and remained that way for a long time (Hofer and Messerli 2006). What brought the floods of 1971 to mind following recollections of those of 1988 was the acute experience of hunger associated with them and the attendant irony that minor flooding brought an entire population to the brink of starvation whereas later major flooding, however inconvenient and destructive of property, did not. Wajid bhai, a middle-aged denizen of Dokhin Teguri, recalled how hungry they had felt in 1971, resorting to cooking and eating banana skins to stave off the most severe hunger cramps only to fall victim to diarrheal cramps, whereas they could much more readily get rice and lentils during the 1988 floods: Back then a day laborer made a few rupees a day and the price of rice was exorbitant, leaving little money for anything else, but now a laborer earning three hundred takas a day can easily afford a kilogram of rice priced at forty-five takas to forty-six takas.

When I began researching this situation attributed by the char dwellers to the immediate aftermath of the 1971 war, I found that this correspondence of flood and famine had not occurred right after independence but a few years later, in 1974. While food supplies were low and prevailing conditions of food procurement and distribution chaotic in the early months of Bangladesh's formation, the situation never reached famine conditions. But in 1974, when food supplies were similarly low, although a rudimentary administration of the "child state" had emerged, floods tipped the situation into a famine. From the historian S. Mahmud Ali's (2010) account of the historical events, it becomes clear that the food crisis was pushed into a famine by geopolitics rather than the floods alone. In early 1974, the low stocks of food grain led Sheikh Mujibur Rahman, the country's new leader, to ask the United States to release food aid promised to Bangladesh. The flooding of the delta and the subsequent destruction of crops soon compounded the situation of low food availability. However, the United States halted its usual food assistance to punish Bangladesh for selling jute sacks, its primary export, to Cuba, which was under an international economic embargo at the time. Mujib went to the United States to plead for aid to be released, but by the time the food arrived, Bangladesh was in the throes of a famine. The famine lasted until 1975, with millions of people reduced to skin and bones (captured vividly in paintings and photographs) and with 30,000 to 100,000 lives lost, predominantly among the rural landless poor.[10]

Although I cannot say with certainty why Wajid bhai and others would recall a famine as immediately following independence when it actually occurred a few years later, this misdating did indicate a possible way to allow themselves to recount a harrowing event that they had all lived through on which national history had remained resolutely silent. It was far less problematic to remember that the famine was the consequence of a flood following a war with West Pakistan and a longer period of economic and political discrimination by West Pakistan against East Pakistan, rather than a famine in which the then-new government of Bangladesh was implicated (Ali 2010). This misremembering entangled the chauras in nationalist historiography and collective memory. Preferring to mull on that experience rather than on the actual year of the famine, Wajid bhai was more preoccupied by the fact that hunger could have had such an intense feeling associated with it whereas now, although people were still hungry, that edge had gone. Rice had been readily available since the green revolution of the late 1970s. Wajid bhai mused, *It fills my belly but it doesn't satisfy my hunger*, a claim I was to hear often.

The examples of these two floods (in 1974 and 1988) show how they were differentiated by the chauras by the specificities of their personal experiences intertwined with national events, with the floods of 1971/1974 recollected as the birth of a nation and a time of hunger, and those of 1988 recollected as a time of countrywide flooding (a rarity until that century), the promise and demise of megaprojects, and the emergence of chars on the satellite maps of the country and the national imagination, along with the emergence of actual char lands for those who had lost their villages to the river in the past.

The floods of 1998 were even more devastating than those of 1988, inundating 68 percent of the country. About 50 percent of the country remained underwater at a height of three meters for up to sixty-seven days, making these floods the longest-lasting of all in twentieth-century Bangladesh (Hofer and Messerli 2006, 313). Yet they did not reach the level of the iconic in chaura memory. The chauras only recalled them as resulting in the highest level to which water had ever risen, making this height the gold standard for raising house plinths to withstand future floods. The 1998 floods' relative unimportance in chaura memory might have to do with the lack of a political dimension to the floods. These were simply floods and not national events.

In comparison to the floods of 1998, chaura memory of the 2007 floods was far more vivid, further underlining my observation that floods become consequential for char dwellers when their experiences are intertwined with national politics, or rather national politics rendered some floods as more consequential than others, coloring char dwellers' memories of those floods. Prior to the 2007 floods, the army had suspended the Bangladesh Nationalist Party's elected government (in 2006), disqualified both the BNP and the Awami League from politics, and set up a caretaker government to oversee Bangladesh's transition to democracy. This government was feared because it seemed remorseless in its efforts to render the country free of corruption, inefficiency, and redundancy, but these motivations were not entirely transparent. The ambiguities of that time were well captured in char dwellers' experiences of the floods during which they found themselves in shelters, a state-provided safe refuge for the duration of the floods, perhaps for the first time in their lives, but they also remembered experiencing more snakebites than in any previous flood times.

Floods and Erosion within the Jamuna River Ecosystem

Floods and erosion belonged together as processes acting on the same topography and within the same ecosystem. If the recollection of floods showed chauras to be Bangladeshi subjects, imbricated in the time and history of the nation-state, what did the experience of erosion have to say about them? To grapple with this question, it is first necessary to develop a point of view in which floods and erosion belong together as part of a larger system.

P. C. Mahalanobis (1927), the Indian statistician whose pioneering studies on floods are still quoted approvingly by environmentalists in Bangladesh, pointed out in the first part of the twentieth century that the floods of what was then North Bengal were a nexus of causes, specifically snowmelt, heavy precipitation, and river flows. This idea of events of nature as constituting a nexus of causes helps draw our attention to the wider context of which floods and erosion are processes, in which they are causative and caused. Or rather, as the Bangladeshi geographer Haroun Er Rashid (1991) has said, our attention should be on the deltaic topography on which the river system is superimposed and which serves as a large drainage area for the subcontinent of Asia. We are asked to imagine not only an enormous landmass draining through Bangladesh but also the surface of land, any surface of land, as overwhelmingly tending toward the act of drawing away water, carrying the suspended load in it (sand, silt, soil), through cracks, crevices, channels, and other pathways to points farther down from it. In other words, surfaces exist in relations of gradation to other surfaces and in constant states of transport and movement relative to one another. And the chauras inserted themselves into this landscape by measuring the high water with their hands, with the precipitous rise of water by two hath (hands) indicating the possibility of floods.

Within this conception of land transporting water and sediment, floods and erosion become part of the constant creep of land and water toward the lowest gradient or the collapse of water-saturated soil into flowing waters. This perspective helps unsettle any notion of land as stable and enduring, shaped by the external action of floods and erosion. Instead, it leads us to think about how land, or rather topography, enacts floods and erosion from within itself. While chars are often treated in nationalist and literary works as aberrations (Baqee 1998), land thrown up and as quickly eroded by the river, we are made to realize that the entire topography of Bangladesh is in effect one large char, the incremental product of sediment brought farther and farther out into the Bay of Bengal by its rivers and ocean tides (H. E. Rashid 1991; Akter et al. 2016). This process is not ancient, as there is little land that qualifies as ancient in these parts. And the low slope of the land testifies to its relatively new origins in geological time, suggesting that water does not flow off it as easily as it would off a duck's back.

Beyond their gradient and surface features and processes, the existence of upraised tracts of land deep under the surface of the delta with depressions and dips alongside them implies that the same topography that enacts floods and erosions from within itself is also conducive to storing water. Thus, at the same time as the topography moves water and sediment to the bay, the tracts and dips within it militate against this tendency, slowing the flows and storing large masses of water. Furthermore, anything that slows or prevents drainage, including technological innovations such as embankments, railways, or roads that create impervious surfaces or cordoned areas with no outlets, adds to the drag on water and sediment. Thus, what we call floods may well be waterlogging, placing it on the same spectrum as bank erosion.[11]

Floods and erosion often play out together but differentially on the landscape. As mentioned at the start of the chapter, floods occur in different temporal cycles, every two, five, or ten years, introducing staggered changes on the topography. Floods that occur frequently are low in intensity and scale, effecting few permanent changes. Those that occur less frequently may overflow banks and transport new material to the adjoining floodplains. Whether this material is useful or not depends on the nature of the material that is deposited, whether it is sand or silt, with silt being fertile whereas sand chokes life, or whether the amount of deposition enhances the landscape or strangles its existing capacities. Floods that occur on a decadal scale result from a greater concert of causes, and they introduce newness in a more persistent way by changing the course and geometry of water channels. They also raise water high so as to raise the height of land. That is, land can only come up to these new heights when water raises it up. The 1988 floods, which bespoke misery for most of the country, were also the moment of the emergence of char land where the villages I studied were located—hence the curious admixture of fear and anticipation that undergirded chaura reactions to major floods.

Floods might structure anticipation and participation in specific ways, but bank erosion, whether the constant creep of land toward water channels or waterlogged land collapsing into flowing waters, only seemed to take away. It might add sediment to water flows and make other possibilities come into play, such as the accretion of land, but these possibilities were pertinent only at some other point in the river's course, either along its other bank or downstream. So erosion seemed to underline one's excision from a place and absence from the present. Consequently, it should not surprise that individual stories, otherwise steeped in flood-related events, were silent on bank erosion. Erosion individualized but did not singularize. One's suffering was simply one's own, with erosion a notch among other such disruptions in one's life narrative. Or, as I claim, erosion pulled the char dwellers into a different state of being and recalling.

Entrainment by Erosion

To place erosion within the everyday is not to say that it was a taken-for-granted element of life, as quickly forgotten as experienced. Instead, it grants erosion the capacity to set the tone and texture of everyday life, for better or worse, rather than to experience it as an external event periodically visited on the everyday. While floods were described in terms of their experienced intensity as *chorom* (extreme), *kothin* (difficult), or *halka* (light), erosion processes were classified in very familiar, even domestic, ways, with the affected soil sometimes juxtaposed to the temper of humans. David K. Wright and colleagues (2013), writing on dust and its entrainment by high-velocity winds along the Middle Gila River in Arizona, explain that those who live in the valley and in the path of such winds have developed complex descriptions of the dust's texture as experienced by their bodies. As I will describe later, chauras too use a highly evolved vocabulary to describe types of erosion, with these descriptions mediating their own physical motions. If in the first instance provided by Wright and colleagues the body mediates the dust's entrainment by wind by speaking of experiences of the wind as making one's body heavy or light, in the instance of char dwellers, language mediates the experience of the body as sediment.

While the term *bhanga* was reserved for the most extreme of erosion processes in which large areas were swiftly broken by the onslaught of river waters, *chapa bhanga*, or the breaking of soil in *chaps* (small portions), indicated the constancy of pressure that soil, and by extension humans, must endure in these parts (Abrar and Azad 2004). In such instances soil broke off as if small pieces of dry bread, just as humans fall apart in small bits through the repeated action of erosion. Such breakage was considered to steal up on one and to happen under cover of night. *Haria bhanga* referred to soil that broke in a sweeping motion as was known to happen to the bottoms of big clay pots or *haris*. In such instances, the swift currents of the water cut into the bank below the surface, leaving the top to overhang. Haria bhanga was considered more dangerous than either bhanga or chapa bhanga. The latter at least gave warnings before their onset through the loud noise or piecemeal action associated with them, whereas an overhang did not give any indication of its crumbling base, as it was in many instances hidden by the river waters. Haria bhanga was endemic to areas around Sirajganj, making people mindful of the eye that suddenly appeared in the river waters, indicating that a pakh, or eddy, had developed underwater, or of bhurbhuri, or bubbles, on the surface of the water as the air previously trapped in the soil was released as the soil broke. Finally, chechra bhanga approximated the sweeping actions of a dishrag on a plate and referred to the sweeping effects of floodwaters on a relatively new, sandy char, as it was swept away by the waters.

Chapa, hari, chechra—everyday words for ordinary objects or actions—give us our first indication of the enmeshment of erosion with everyday life. A second indication was that chauras often said they were bhasha tana lok (people who float and are pulled along or people who float and pull [their houses]). This comment might be taken either as a statement on political marginalization and economic deprivation or as an expression of their autochthonous way of life. Their marginalization within the Bangladeshi state and society was obvious, as presented in the introduction to this book. But claims of chaura autochthony were dubious at best. As mentioned in the introduction, the Jamuna River had been flowing in this southern direction since only 1830 and with this amount of force since 1950. The people who constituted the chauras at present had only been in the throes of its motions since around 1950. And because the river was braided, with channel avulsions along a west-to-east axis, it was not clear whether the chauras remained char dwellers permanently. When a river channel moved east or west, it left behind people who now lived on land that had become permanent and that would remain as such unless the channel moved back. Entirely new groups of people on the river's path became the chauras as they experienced erosion, perhaps for the first time in their lives. Instead of thinking of chauras as comprising a distinct culture, it is more useful to consider the topography, or more specifically erosion, as temporarily entraining those who sheltered on chars into this way of life. If, as I have been suggesting, we are to think of erosion as setting the tone and texture of the everyday, one way to understand this is to think of erosion as the means by which the river entrains the everyday. This means taking seriously the chaura claim to being people who float or are pulled along not only as an expression of their vulnerability to the world but also in the sense of being entrained within the flows of the water and the breakage of land.

There are several distinct senses in which this entraining might take place. As mentioned earlier, in geomorphology, entraining indicates the suspension, channelization, and deposition of sediment within the currents and flows of water. We might take the chauras to mean a similar suspension within the whole, much like sediment in water flows. Physical entraining is not unthinkable for social beings. For instance, human diurnality already indicates biological immersion or entraining within the cycle of day and night. Within musicology and social psychology (Clayton, Sager, and Will 2005), entraining has acquired another aspect as immersion within the rhythms of another person or bringing one's rhythms in alignment with another's. We are already oriented toward entraining as a process within intersubjective relations. Drawing on the cybernetics-influenced anthropology of Gregory Bateson (1972) that sees homologies in pattern formation across different forms of life, even mutual replication across different domains, I would extend an understanding of entraining to include replication. In other words, the char dwellers were not only suspended, channelized, and deposited by the river; they also imitated the river's tendencies and replicated its patterns within their everyday. Consequently, with erosion within the everyday, entraining by erosion entailed being as sediment in the river waters in both receptive and active senses.

Chaura Self-Opacity

Erosion also induced being obscure to oneself. Chaura self-opacity was made evident to me through their recollections of the events of erosion that they had experienced in the past. Early in my fieldwork I realized the extent of mobility packed into people's lives. I decided to carry out an exercise of tracking individual movements by providing each person with whom I had a conversation with a blank map of the area. The map had only the schema of *mauzas*, or political units, overlaid on a satellite image of land and river formations of the area. I asked them to indicate on the map each move they had made over the course of their lives, what brought it about, why they moved to where they did, who went with them, and any other information they considered necessary to flesh out these moves. I tried to get a spatial spread by including those who lived on the mainland shores closest to the island char on which I resided, in addition to those who lived in the three study villages within the char. But other than incorporating this spread, the selection criteria were open, inclusive of the elderly and the young, women and men. I ended up with fifty such maps from people aged from their mid-twenties to their nineties, for an average age of fifty-five, with more male respondents than females. A short interview spelling out the details pertaining to the moves plotted on the maps accompanied each map. Fairly quickly I decided not to guide this interview through prompts or questions. My intent was not to generate the most accurate accounts but to see what bubbled up as the most consequential elements in life narratives.

It became clear from the outset that women's recollections were different from men's. While men gave definite answers, laying out their experiences in a recognizable chronology from the time they were children to the present, women tended to get very confused in terms of chronology, speaking of events of recent movements as if they were from the distant past and recalling those from their past as if they had just happened. The crowd that would invariably gather around us would heckle the women, saying that they didn't know anything, or else women would demur by saying, *Oto ki mone thakey?* (Can so much be remembered?). Some of them deferred to their husbands or children. Yet in the act of inscribing their accounts, I realized that there was great consistency among women's narratives, and that more often than not the appearance of confusion arose from the fact that they experienced their marriage and subsequent absorption into their husbands' lineages and families as disruptive events equaling those of erosion. In other words, when asked to recollect when they first moved, they were as likely to recall their marriage as they were an actual experience of bhangon, whereas men more often remembered having picked up wives and dropped off sisters and daughters along a journey largely punctuated by bhangon.

Morium, one of my most cherished friends and interlocutors among the chauras, a woman in her fifties, provided perhaps one of the most vivid descriptions of bhangon but one that was completely entangled with the story of her marriage at a very young age to a poor man who would continue to disappoint her until his sudden death some six years ago, leaving her bereft, abandoned by her three grown children and responsible for the care of her youngest daughter. She could not express her criticism of her husband in a direct way, but her narrative of erosion provided ample clues to her embarrassment over him, as she described erosion as an event that tested their individual mettle and during which her husband proved to be wanting. She related:

> I had only been married a few years. You remember, I was only sixteen when I got married to someone of my own village. Why would my parents

give me in marriage to such a poor man? I quickly had two young boys and a baby girl in my lap. The floods came and we took refuge in the voting center, but the village elders would not let us stay for long. We had to find a place of refuge. My grandfather had fields that he cultivated that were somewhat remote but dry. He put us up there. The waters came. It made such a noise that we woke up screaming and shouting, huddling together on the bed as everything else that we owned floated away. We heard land groaning and falling into the water. I could not stop my screams. My husband, what could he do, we did not own a boat. We just stayed like that that night, the longest night of my life. Even a snake took refuge with us, coiled up on the headboard of our bed staring at us while we stared, terrified, at it. My husband spoke to the snake, saying, "We will leave you in peace if you leave us in peace," and that is what happened that night. The next day my *dewor* [husband's younger brother] came to get us on a boat. They had heard our screams but had to wait until the first hint of light to come rescue us. My dewor, he killed the snake right away, my husband couldn't, but my dewor did. After that we went from here [Rihayi Kawliya] to the other shore [Khas Pukuriya], where many of our neighbors had taken refuge. There we stayed on the behest of Montu Chairman. And then I had the heart condition. I would work just a little too hard, and my head would swim. My sons and daughters [presumably by this time she had had her second daughter, and her children were relatively grown up] shouted and shrieked to my husband that I would not live much longer. He shook his head and went to the fields. What could he do? He was a day laborer. It was my parents who gave me in marriage to him, so how was this his fault? On the day he died, I was lying in bed. I had had to be revived with buckets of water poured over my head. The children waited impatiently for their father to come home, asking every so often, "Where is he? Shouldn't he be home by now?" He had come back but just as he came to the bottom of our household, he fell down and probably died right there and then. Here we were standing on the top of our household, looking into the distance waiting for him to come back, and there he was lying dead at the base of the household. What was I to do? I tied a cloth around my aching head and headed to my brothers. I begged them for help as my sons had married by this time and they did not want to be responsible for me, much less for my youngest, Shilpi. I couldn't just abandon her to make sure that I get daily food. I just went into the darkness with my youngest.

In saying that she walked into the darkness, Morium was referring to the fact that her brothers gave her a patch of land on the char that had reemerged in Rihayi Kawliya, from where she had first escaped floodwaters and land erosion. As a new char, it was as yet relatively uninhabited, and she had to clear catkin grass, cut earth to raise land for a new household, harvest *badam* (peanut) and *kaun* (cereal) to sell for food—that is, do everything anew in a village where she had been born and raised, gotten married, and become a mother but which was now a wild frontier to her, and she its settler.

This conflation of marriage with erosion-induced disruption introduced another break into women's accounts. In the process of dotting maps to show their movements, they would invariably stop at the point at which they got married. At this time, each woman would ask her husband, usually sitting next to her, which movement to track after the marriage, hers with him or that of her natal family. While this question would garner the exasperated response that, of course, it was only her movement with her husband and his wider family that mattered after this point, this inevitable pause and reflection showed that while marriage conjoined a woman's fate with that of her husband, she retained a connection to her natal home and a shared sense of the disruption and dispersal brought on by erosion and other natural disasters in the lives of her parents, brothers, and sisters. In other words, women were attentive not only to their own trajectories marked through their husbands but also to those of their intimate others, putting their movements in constant relation to their natal families. After all, they might have to catch up with their family wherever they had gone—sometimes quickly, as in the case of Morium with respect to her brothers.

While the past was recounted in this very enfolded manner, the present was often a site of absence. Several women floundered as they tried to provide an explanation for why they were here and how they had gotten here, again deferring to their male guardians. They simply said, *Disha haraye felsi* (I have lost my sense of direction/trace of the path). Those who were heads of their own households, such as a widow like Morium, were better able to locate themselves in the present but seemed to ceaselessly worry about what they would do next and when they would come to be without options. This relationship to the past, the constant tracking of the movements of others along other planes of erosion and the frequent evacuation of the present pointed to the fact that while erosion might be the occasion for bravery, heroism, courage, or even endurance, it left its deleterious mark on one's sense of space and time.

In contrast, as mentioned earlier, men much more readily conveyed a clear sequence of moves they had made since first experiencing the disruption of erosion. Even if they had not yet gained the capacity for understanding, rendered as *budhi geyan khola* (the unlocking of intelligence and knowledge), at the time of their movement, they knew whom they went with and where they went through listening to the stories of their fathers. As sons, brothers, fathers, uncles, and grandfathers, they seemed to be the keepers of the memories of their families' trajectories. They recollected that they all went together. They recounted that their family moved just a bit east closer to the big banyan tree or next to the site of the *hat*, or weekly market. They remembered that a big landlord in the area who was known to them had given them refuge or that a chairman of the subdistrict had given them the unofficial nod to occupy the embankment, which was on government-owned land but had once belonged to the chairman, or that the occupants of the now destroyed village had all gone together and forcibly set up homes in an empty field in some town on the mainland. They started these journeys in collectives, as a village, a group of related households, members of the same gusti, or clan, selecting a matbor from among them to negotiate their passage to another place as they moved from one site to another, either trying to stay just a little ahead of the river so as to be able to be within eyeshot of the lands they had lost and hoped to regain or making a wide arc from their original site so as to put themselves out of the path of the river. But they invariably reached a point when they could no longer move together. Either they lost the will, the bargaining power, or the material means to stay together or else there was nowhere for them to go, leading each to seek out his own path, rendered as *je jeye bhabe pare* (however each could manage it). This could be the point at which fathers might separate sons from themselves, a process that happened much later among the chauras than in the wider Bengali rural populace, so that each might (re)start unencumbered by numerous family members, or when elderly parents were deposited with the few members of the family who had found stable ground deep within the mainland, in Rajshahi, Tangail, or Dhaka, the most frequent places to which "erosion-induced displacees" migrated. This tapering of the corporate group was fairly consistent across the movement maps I collected from men.

Just when I thought I had grasped the stages of the movements produced by erosion, I realized that the men were also guilty of elision, although of a very different kind than that of the women. If women tended to forget sequences of moves, to loop back to those moves that had been particularly searing for them, or to get entangled with the movements of their natal families, men were forgetting or deleting those that took them more and more inward before finally giving up on a place as a lost cause and moving out to seek shelter elsewhere. To explain further, when a house first broke, particularly in the case of chapa bhanga, or piecemeal erosion, the family involved preferred to take refuge in the village, on the edges of another's household, rather than move out entirely. They might simply prop up the tin roof of their house within someone else's household or arable land, preferably a relative's, a mode of habitation referred to as living *aangaa*. Such arrangements might continue indefinitely as people preferred these deprivations to leaving familial ground and setting up home among strangers. If the sheltering house also fell victim to erosion, both families might take shelter in the household of a third and so on until all were piled up on top of one another holding on to the last bit of the village. Such arrangements were quite common and seemed to be accepted as par for the course. However, they were also generative of great tensions among households, with families torn asunder by conflicts over toilet arrangements and spats between children.

These movements inward produced dense whorls when rendered as lines on my movement maps. Men decried their confusing effect on the clear trajectories they strove to present, deleting them as they went along so as not to confuse their narratives, to focus on the main ones or so they said to me. But what they were leaving out were the feelings of betrayal that inevitably followed upon the movements out of the village. On numerous occasions a family pointed out another family that had lived with them for years on end when the need occasioned but who did not even lift their eyes in acknowledgment of the first family when this family had need of them.

In time I learned to elicit these elisions by changing my question from "When/where did you move?" to "When/where did you pull your house?" as the transplantation of the house materialized chaura movements more consistently within the maps. I started thinking of the maps less as records of people's movements across space and time and more as records of their splintering, forgetting, and elisions, what one might call their lack of understanding of themselves. Yet another way to understand this relation to oneself is to understand it as an effect of the thickness of one's relationality with others. As Judith Butler writes, "The opacity of the subject may be a consequence of its being conceived as a relational being" (2005, 20). I take chaura self-opacity less as an indication of their incomprehensibility to themselves, and more as an indication of the thickness of their ties to others and to the larger movements within which they were immersed. Chaura self-opacity was a trace of nature as the unconscious within them.

Moving Lands, Shifting Grounds, and the Birth of Morality

It was the movement maps that also helped me understand the moral significance of the material house, moth-eaten, decayed, hardly worth the effort of carrying, but still the means with which to step into the world. The picking up, the pulling or dragging, and the resetting of the house, more than simply a roof over one's head, was the means by which the chauras enacted their selfhood of both dependence on others and self-sufficiency, which we might take as another way to understand what it is to be acted upon. Another means by which those displaced for the first time propagated their movement was by buying land and moving their house onto it and trying to do so as often as they could until the point came when they were simply grateful for any land on which to take refuge. But at some point along the trajectory, when all material means had disappeared, one reached for the hand of another as the means by which to propel oneself forward. Recall Naznin in the scene of erosion who had caught the hand of another in her desperation to be able to move out of the village of Boylakandi to Nauhata, when her orphan status and her illness had left her with no options.

In an interesting reversal of Bengali kinship norms (Inden and Nicholas 2005), instead of being burdens, women became the most valuable networking asset within families as they proliferated the lines of escape and exit from increasingly impossible situations. Women, married off when young, led their affinal families back to their natal homes; mothers mediated relations to maternal grandfathers and uncles who were more likely to give the mother's rightful inheritance to grandsons and nephews than to their daughters and sisters; sisters and daughters were deposited like silt along journeys but remained alert to their original families' changing situation. Women also welcomed natal families back, giving unusually expansive access to their in-laws as additional folks from whom to seek help in times of need. People described impossibly faint lines of connection as their lifelines out of erosion. These kinship relations, faint in some cases and fictive in others, ran the risk of accruing more dependents than when one first started off, with families carrying with them in-laws who had once generously housed them and whom they were subsequently obliged to take with them when their land returned or the elders' land broke. A young man once spoke teasingly of a paternal aunt who had moved away to the mainland, a place of escape to his mind, but who had foolishly chosen to return to the char when her brother's land returned and he returned with it. The young man said: *We kick them but they still stay around.*

Many, including those who had no land of their own to lose in the first instance, waited for their villages to return so that they might return with them. There were some obvious lines of continuity. For instance, if one's leased land went into the water during the period of the lease, the owner was obligated to allow the lessee to complete the lease if and when the land returned. To live among those who knew one and with whom one had standing (*sthayi, thikana, ostitto*) carried more obvious weight than living among strangers. Even so, certain villages have a better reputation than others for both taking back kin and taking in strangers. Villages such as Dokhin Teguri were known to be *baro-mishali*, or twelve mixtures, on account of their absorptive capacities whereas others were under the dominion of a few. This was the point at which we entered the scene of erosion at the start of the chapter. Five bighas of land returned in Boylakandi had enabled eleven households to be secure for ten years before the land broke again, revealing anew the importance of such numbers and calculations that enabled chaura existence.

There have been decadal changes to the absorptive and reabsorptive capacities of the area, as I gleaned from the movement maps. When the river first started moving and erosion began on a large scale in these parts in the 1950s and 1960s, entire villages were sanguine in their expectation that they would be able to find refuge in other people's homes, on government property, or on unused land lying about somewhere. The Badh, or the embankment on the western side of the Jamuna, was home to many, even as parts of it crumbled and new parts were put up farther along. Even now there are neighborhoods along those parts of the embankment that the government had abandoned and declared unsafe, vulnerable to further erosion. As the squeeze of land continued, chauras expressed their desperation for refuge by taking up living alongside roads, perhaps the most precarious kind of refuge of all, although not without its benefits (Indra 2000). By the 1970s, as war broke out and Hindus left the country in droves for India, their land became available for sale or occupation. Consequently, char dwellers' narratives indicated a marked increase in land sales and possibly the emergence of a market for land underwater, as several of my acquaintances took to engaging in such sales either to manage lean years during and after the war or to speculate on a future of lands returned by the river (see chapter 1). In the 1980s and 1990s, as the Bangladeshi government turned its attention from postwar rehabilitation to the issue of rural poverty and unemployment to produce incentives to keep people from migrating into cities (S. M. Ali 2010), the Bangladeshi countryside was opened to more market forces than before, which was a mixed blessing. In the 2010s, a somewhat different scenario predominated. As a result of the green revolution, more land was needed for *irri* rice cultivation, with pressure falling on char lands. This produced the lure to grow more rice for greater productivity and profits. Within this scenario it became increasingly difficult to move as a corporate group as landowners were reluctant to give up productive arable land to provide refuge to those displaced by erosion.

The char dwellers who traveled these paths were clearly brave, physically capable, and alert to every opportunity. But they were also vulnerable to deeply unequal relations, chronic uncertainty, the betrayal of intimates, the possibility of growing indifference toward their suffering, and the exhaustion of their capacities to carry on. Researchers Abrar and Azad received the following response to their question as to why people stay in what feels like a situation with impossible odds: "They pin their hopes against hope that erosion will not affect their own household. They were scornful that some people move away with the slightest hint of a disruption. In their eyes, they were not respectful enough to the homes and hearths of their ancestors. They claimed that they were 'not *bhatias*' like those from Netrokona or Mymensingh, 'we have much deeper respect for our ancestral homes'" (2004, 47). Yet immersion in this flow extracted its dues. An elderly man in Dokhin Teguri complained to me that he was no longer human. In his words, *I am now the shell of a crab [kakrar khosha]. You know about the crab. It dies after giving birth that one and only time and leaves behind its shell. I am that shell.* These words gave vivid expression to his sense that he had a skeletal existence, he lived without the spring of life, but at the same time his was an ambivalent expression. On one level, it spoke to the denudation of existence through repeated experiences of erosion. At the same time, becoming a shell had happened only in the aftermath of giving birth. What was it to which this chaura saw himself as giving birth before becoming merely a shell?

To open this question up for further deliberation, I loop back to an earlier claim that the chaura everyday manifested entraining by the river system mediated by bank erosion. Another way to put this is that nature expressed itself most forcefully through the chauras, particularly their experience and recollection of erosion. Erosion put the chauras in the waters and delivered them up to processes within flows and currents, those of absorption and sedimentation, a meander here, an avulsion there. The chauras experienced their lives as floating and being pulled. I took this expression of theirs to be perhaps the most obvious trace of their perception of entraining. Other expressions, such as the detailed descriptions of types of erosion, indicated a replicative relation between their movements and those of the river. But there were even further expressions that metaphorically resonated with their impressions of the river. To arrive en masse upon the shores of a proximate village was described as coming like a *dheu*, or large wave. To straggle in, a few families at a time, was described as coming like a *shakha*, or snaky branch, of the river. These resonances with the river were also evoked in justifications of new social norms, and not just of physical movements alone. To return to one's parents' home or to seek refuge with one's sister or daughter was not viewed within the usual frames of reference of *lojja*, or shame. A common refrain was, *How could there be any shame in returning to one's mother's lap or of reclaiming loved ones lost along one's journey if the river does the same?* This expressed the commonly held belief that the river's many strands were merely lost and wandering in search of their own beloveds to be reunited with them one day. These parallels between the river's and one's own movements and actions certainly indexed the impingement of the landscape on chaura lives. But it was the inexplicable force of the obligation to give refuge or the forcefulness of the demand to be given refuge that could not be fully assimilated into village belonging, kinship relations, or political patronage that marked morality as the product and compulsion of the landscape, as the completeness which impersonal nature gave to human acts.

Conclusion

While Bangladesh has the reputation of being a flood-prone country, this chapter has described how land erosion is as frequent and destructive as flooding, if not more so, at the register of sociality and the individual psyche. The importance given to floods both internationally and nationally meant that they were much more tied to the state and people's imagination and punctuated the country's history, with some flood events more iconic than others. The chapter has explored how char dwellers showed themselves to be Bangladeshi nationals and historical subjects by weaving their life narratives with those of floods.

Erosion operated in a different way than floods in the lives of chauras. Erosion was more destructive and world annihilating but was experienced in a piecemeal fashion, with one individual and then another, one family and then another, one village and then another coming under erosion. Yet concerted attention to chaura experiences of erosion showed how erosion set the tone and texture of chaura everyday life, self-descriptions, and life trajectories. Moreover, chauras inhabited a different state of mind while in the throes of erosion, described as being dependent, other to themselves, or like objects plied by the river's waters. And this condition, produced by what I call entraining by the river through the mediation of erosion, was made manifestly clear in their recounting of their experiences of erosion that lacked the detail, specificity, and poignancy of their remembrances of floods.

When one recalls that the river moved very often such that those who lived on chars had only come to do so recently, at most one to two generations ago, while those who were previously living on chars either had their land back permanently or had lost it for good, several things become clear. Chauras did not have any transmitted wisdom or habitus of what to do in the event their lives went into the water, but they somehow did the right things. They had little or no recall of what they did because once erosion started, it was almost continuously a part of their perforated lives, therefore apparently relegated to the opacity that overcomes the most familiar.

Erosion activated an entire skein of capacitation that could only be so if it were a kind of noncognized nonacting. I speculate that this unconscious existence is the entrainment of the chauras by the river, and the obscurity surrounding its details and the morality that was expressed and upheld during erosion the aspect of nature within them. It showed how the chauras perdured within the char-river landscape as an element within it, with their intelligences in relation to the intelligences within the landscape. And I find this to be a very persuasive account of the romantic understanding of passive receptivity about which Dalia Nassar says: "The romantics repeatedly argued that the possibility of an ethical life depended on an original unity between mind and nature. This did not simply mean that human beings should be able to transform the world with a spontaneous and free will, but also—and in some cases more significantly—that the human being should be *affected and thus transformed by the world*" (Nassar 2014, 4).

Shohidul and the author located
Nojrul Mistry atop a building
that he was breaking down into its
parts. A young boy is visible,
undoing the house from within.
Photo by author.

Young boys run as they carry the siding of a house to the boat to be transported, spurred by the promise of some money and a hot meal. Photo by author.

An old man keeps an eye on the
goods being gathered for transport
but almost looks like a bundle
to be transported. Photo by author.

The weaver finches were about to experience their first erosion event, with their nests destroyed as their tree home tipped into the water. Photo by author.

3.

Elections on
Sandbars and
the Remembered
Village

In the Mind's Eye

I WAS IN THE VILLAGE OF DOKHIN TEGURI in Gorjan Union, Chauhali upazila in Sirajganj district during the 2011 local elections, whose purpose was to elect representatives of groups of villages or wards to the local or union parishad level of government overseen by a chairperson. This chairperson in turn represented the union parishad at the upazila or subdistrict level, and the chairperson at the head of the subdistrict level of government then represented the upazilas at the zila, or district, level. A minister would finally break into the national scene by representing one of the six constituencies to represent Sirajganj district within the Rajshahi division in the Jatiyo Sangshad, or the national parliament. The elections I witnessed were for the union parishad and the upazila level—in other words, two tiers up from where I was physically located for fieldwork in Dokhin Teguri. These local elections came after the national elections held in 2008 under the military-sponsored caretaker government.

Initially I was frustrated with chauras' absorption in the elections insofar as it made even more uncertain the hours they were to be found at their homes. It made preambles to interviews sometimes overtake interviews entirely as interviewees mulled over every detail of the elections with

Shohidul, my equally election-obsessed research assistant. Election preparations seemed to absent them from the materiality of their environment, which was my point of entry into chaura lives. However, being in the midst of the election preparations, watching the elections happen and then later seeing their effects roll out over time, I realized that they provided a unique vantage point from which to understand the chauras' modes of grasping their situation.

By awareness of their situation I mean several things. The election I witnessed was part of a larger context of electoral politics and its administration in Bangladesh. Within such a context, awareness of one's situation could mean the extent to which this election was seen to be in service of a larger political process, be it democracy or otherwise. The product of colonial and postcolonial histories aimed at securing local self-government, the election helped to trace out the changing figurations and fortunes of the quintessential village in rural Bengal. The election also brought into focus the status of the village in more recent neoliberal imaginations of self-governance. From this vantage, the char dwellers' awareness of their situation might also be taken to be an awareness of how their village compared to the generic village structure that undergirded the economy, sociality, and polity of Bangladesh, which was over 70 percent rural.

The first indication as to why these elections must be about a different setup and awareness of it, and not just the overlay of political administration on villages, came from the fact that quite a few of the elections that were held in these parts were for villages that no longer existed. In the struggle to ensure that these elections were pulled off, I saw long-scattered villagers traverse pathways that ran counter to the river's movements, which previously had washed away their villages. They gathered together so as to make their villages present and able to be counted. And no sooner did they do so than they returned to their present places of residence to wait and watch the effects of their actions. They also watched their newly elected politicians be incorporated within the administration through their fitful efforts at fulfilling their official charges and the more expected efforts at filling their own coffers through expropriation. Both sets of actions produced effects within the river-char landscape and the Jamuna ecosystem of which it was a part, inclusive of the monsoon winds and rains, the angles and gradients of the surface, the movement of water and sediment, and the tracts and dips in the subsurface and below, which roughly correspond to the dimension of Bangladesh as a delta (see chapter 2). And these actions, whether official or corrupt, were precisely what the chauras had long envisaged in their imagination and through their strenuous efforts at participating in elections.

This chapter explores the specificity of char-based elections within the context of national elections. It examines the colonial and postcolonial history of self-government and elicits the nature of chaura participation in local elections as an expression of self-government. The historical, legal, and emotional imbrication of one's village of origin, and landholdings therein, with elections and self-government gives us one set of rationales for chaura participation in elections for lost villages. Their modes of participation give us another, as these were the means of chauras' reembedding within the Jamuna ecosystem after their excision from it by erosion. Finally, the chapter explores chauras' reconstruction of their situation, through their imagination and in narratives that weave together the past, present, and possible futures to show how they further stitched themselves back into the landscape and ecosystem, through thinking of themselves as vital to its internal workings. I draw on Schelling's mentor Goethe's perspectives on the power of the imagination to not just replay and appreciate the workings of nature in the mind's eye but to allow char dwellers to insert themselves and participate within nature. As Goethe writes, "At first I will tend to think in terms of steps, but nature leaves no gaps, and thus, in the end, I will have to see this progression of uninterrupted activity as a whole. I can do so by dissolving the particular without destroying the impression itself. Rough separation into dynamic elements. Attempt to refine this. Attempt to discover further intermediate points. If we imagine the outcome of these attempts, we will see that empirical observation finally ceases, intuitive perception of the developing organism begins, and the idea is brought to expression in the end" (Goethe 1988a, 75). And further, "Why should it not also hold true in the intellectual area that through an intuitive perception of eternally creative nature we may become worthy of participating spiritually in its creative process" (Goethe 1988b, 31). Goethe's perspective on the power of the imagination helps us to see the diversely ramifying potential within chauras' construction of the Jamuna system in their imagination, in their journeying to congregate to vote, and in their long-term projections of what might result from the elections. Chaura construction of the river's movements with their own movements folded within those of the river helps us appreciate how a village and its elections might be a part of nature as much as a sociopolitical process.

Elections in the National Context

The 2008 national elections were widely hailed as one of the most successful elections in the history of Bangladesh in terms of the low incidence of violence and wide voter turnout (Nizam Ahmed 2011). A very different scene greeted me in January 2014 as I watched the unfolding of what was being openly called in the national and international media "a farce" of a national election. Sheikh Hasina, the leader of the Awami League (AL), who had been elected as prime minister in 2008 under the supervision of a caretaker government, had the constitutional provision for such a caretaker government removed. Whatever her motivations, her action led to the fear that elections without a caretaker government would not be fair, the boycott of the elections by opposition parties, months of enforced travel blockades and work strikes, and an open season of violence by the standing government and opposition parties against the general populace.

This political crisis produced untold misery in the chars. Their state suggested that while chauras suffered the weather and stood to suffer the climate, they might have been better equipped to deal with these exigencies than they were with prolonged restrictions on their movement by the political crisis, as it was short-term migrations that enabled life on chars, where lean seasons predominated and food was not a guarantee year-round. But what made the elections, conducted with grim determination by the Sheikh Hasina government, a farce was that despite the fact that voter turnout was low, both because many seats were uncontested and because of the threat of violence, and that all that the government had to do was to simply hold the elections according to the mandate it had given itself, they still managed to show "robust" voter participation. The elections rode in on a crest of vote rigging.

The political scientist Ali Riaz (2014) declared the 2014 elections just as they should be in Bangladesh, in which elections were held not to put into effect public-minded politicians and policies but to win a popular mandate to allow the head of the government and the majority party to continue to do as they wished, leaving the rest of the population to their misery.[1] However, as Mushtaq Khan counters, such elections have not produced a straightforward oligarchy with those in power dominating all economic opportunities. Rather, the presence of factions within the political system, by which individuals endowed with authority attract and maintain allies and subordinates across class lines, have meant that elections occasion jockeying for positions of power even within the head of government's cabinet and political party. Even small shifts in position have ramifications through-out the ranks of factions, providing or removing opportunities quite a ways down into the grassroots (M. H. Khan 2000). This perspective allows us to understand why local-level elections might be significant in providing oc-casions for different factions to enter the administrative hierarchy.

Yet such prosaic understandings of elections do not adequately explain the enthusiasm with which men and women, many of whom bore no appar-ent relation to dominant factions, partook in either the 2008 national elec-tions or the 2011 local elections that preceded the 2014 farce. One scholarly perspective views elections as spectacles that draw in people through the lure of entertainment but whose primary effect is to enact the standing political order, much like a court ritual (Banerjee 2007, 2011). Another emphasizes the aspirational aspect of participating in elections by which the poor and the marginalized express their hopes for the state (Ahuja and Chhibber 2012). Amid this seemingly irreconcilable field of perspectives, David Gilmartin (2012) has suggested that specific accounts of elections can draw profitably from classical concerns on influence peddling and the (in)capacity for self-sovereignty or decision-making within what he calls the global history of voting. While this way of posing the problem may make it seem as though we are being asked to reify autonomous individuals and free, unfettered choice, I read Gilmartin as asking us to consider what we mean by influence and how we might take voting to attempt sovereign acts even within distinct fields of influence. If previously in the book I have considered the productiv-ity of being imbricated in kinship and property relations and the productivity of being opaque to oneself in one's experience of erosion, here I consider the productivity of being fettered within local electoral processes and practices. With this provocation in mind, I consider the widest field of influences en-compassing the 2011 local elections in Chauhali, Sirajganj.

In the Grip of Election Fever

Initially, as mentioned earlier, I was indifferent to the electoral preparations going on around me as I tried to keep interlocutors focused on the phenom-enology of weather and the moving ground beneath their feet. As the swirl of energy around me intensified, I too became gripped by election fever. One late evening in June 2011, I was walking home with Shohidul from the Nauhata Bazaar, having closed up shop with Mukhtar bhai, the owner of the most popular tea stall. Even after the unusual election-time busyness of the market had ceased and the last generator had sputtered out, we kept running into groups of people. I could see faces very clearly for a change, as it was a moonlit night, making it easy for us to pick our way home on spontaneous paths past jute fields and through eucalyptus tree plantations. Unusually but not unexpectedly, these people were not from around here; some were from other villages on the island, and some were from villages and towns even farther away, across the waters to the east and west of the char on which Dokhin Teguri was located. I knew some of the men (at this hour it would only be men moving about with such ease), some of them knew Shohidul, and they stopped to talk to us. Because neither Shohidul nor I could vote here, I took their intent as simply seeking to share.

The men spoke of their candidates with tremendous pathos, describing how their voices had gone hoarse from exhorting people to vote, how their feet bled from walking these pathways to reach every village, how they risked slipping into the water slumped over exhaustedly in slips of boats that took them from shore to shore, how some boats did not appear when needed, in which case there was nothing for the candidate to do but to dive in and swim the distance. Some went into the details of a long-ago event that had estranged a village from a particular candidate, trying to present the candidate's feelings of hurt and desire for reunion with the villagers. Yet, fearing for their skins, these candidates could only send their representatives. I was taken by the melodrama.

There was also an inexorable quality to public participation in election preparations. Just as these bands of electioneers roamed from village to village to convince their denizens to vote for their candidates, villagers stayed up much later than usual, allowing these outsiders to enter their homes. They sometimes kept a stack of betel leaves and nuts that they could least afford to give to the visitors to chew while the latter exhorted and advised. Children took to marching around the village shouting out slogans in support of a particular candidate while sucking on sweetmeats that had been offered as enticements: "My brother! Your brother! Nojrul Brother!" People took evident relish in the exchange, with women rushing out of their homes to speak to those in the know who might be passing by. And why shouldn't they, Shohidul explained. Bangladesh hadn't had local elections since 2003, eight long years ago, during which the rise and fall of elected governments, military coups, and caretaker governments had intervened. And, despite their insufficiencies in meeting the needs of their constituencies, not to mention the corruption structured into the system, local elections were known to grow leaders for national politics (Westergaard and Hossain 2005; Siddiqui 2005). The then minister of fisheries and livestock, Latif Biswas, for instance, was Sirajganj District's very own.[2]

Let me pause here to give an account of one of the more hotly contested elections for chairpersons in the char to give a flavor of their personalized nature and to elicit a possible field of influences within which chaura voters were immersed. Contemporary Bangladeshi politics was largely organized by party affiliation, with party organization extending deep into the city and countryside (Lewis and Hossain 2008; Shehabuddin 2008; Ali 2010). Both the AL and the Bangladesh Nationalist Party (BNP) had local representatives in Dokhin Teguri who were integrally involved in regional party structures that were themselves subsets of larger all-country formations. So, for instance, local representatives of the AL were on the ready to stuff ballot boxes on the eve of the 2014 farce while local representatives of the BNP were waiting by the riverside for the arrival of the boxes in order to steal them before they could be used, all by the dictate of "central," as both party headquarters were called.

Despite the clear dominance of party politics, I was repeatedly told that party affiliation took second place to the choritro, or character, of candidates at this level of politics. It mattered more to have someone who was known to be good rather than someone who was known to be otherwise, no matter how well backed he or she might be. At the same time, any period of involvement in public life and governmental office tarnished one's integrity. The common presumption was that all politicians were necessarily corrupt. Because government salaries were low in comparison to the cost of electioneering, it was widely understood that the shortfall was made up through illegal extraction from local revenue collection and national government–financed projects (see van Schendel 1981). Consequently, character was being judged by criteria other than that of financial honesty.

In further specifying the worthiness of political candidates, it should be noted that while the figure of the matbor had usually predominated as a leader within Bangladeshi village structures, and this position had tended to be tied to land, family (poribar), and clan (gusti), the electoral system had diffused, if not disrupted, the relationship between land, lineage, and authority. Many matbors were now village leaders on account of having been elected to a government position and retaining a claim on that title long after their tenure ended. They might subsequently acquire land and improve their lineage, thus blurring the lines between traditional leaders and elected ones even further. This suggested a high degree of state penetration into village life and political mobility (S. Rahman 2006).

That midsummer in 2011, five men were contesting the position of the chairperson for Sthal Union, which was made up of wards comprising villages just north of Dokhin Teguri. These men were well known to the villagers. Nojrul Chairman was the incumbent who had been in the position off and on for the past two decades, a few times through reelection, other times through delayed elections as in this last time.[3] He was well liked although there was a growing feeling that there was a need for new blood. His competitor was Hatim Master, a high school teacher employed at a government school on the mainland who sought to represent the chars. Living at a distance from the chars had been held against him in the past, and the last time he ran he lost to Nojrul Chairman. It was now felt that having a representative on the mainland might help to draw the state's attention to the chars.

The third competitor was the infamous Ayan Darogah, so called for being a police inspector, who was known to have been the mastermind behind the murder of a chairman named Mian Ullah in the late 1980s but for which he was never charged. He was back on the ballot this year. However, his involvement with this murder kept surfacing, thus making it difficult to conceive of him as *manush*, or human, much less *joggo*, or capable of having a good character, although ultimately he did do well.

The two other candidates were the official choices of the political parties, but they seemed to pose no serious challenge to the incumbent and his main contenders, Hatim Master and Ayan Darogah. The BNP candidate was widely maligned even by his party members for having bought his ticket to stand for election from the BNP party chairman. It was felt that this bribe money could have been better spent in campaigning—that is, by distributing it locally rather than giving it to only one person. The AL's official candidate seemed intent on publicly humiliating Nojrul Chairman, claiming that his own funds had been raised from among those who had sold their votes to Nojrul, who was thus thanked for indirectly paying for the AL candidate's campaign run. This was a scoffing reference to the large sums of money, to the tune of BDT20 lakhs (approximately USD26,000), that Nojrul had already spent on the elections, matched only by Hatim Master. It was repeatedly mentioned that Nojrul Chairman was being bankrolled by his seven brothers who were established at jobs and businesses outside of the chars and who bid him to hold the field for as long as possible with the promise that they would saturate the field with money a few days before elections. Hatim Master also had well-established family members whom he had set up in government jobs and who were helping him. More important, he had the moral support of his relatives who lived in one of the largest villages in the char and whose loyalty was unquestioned, again a mocking reference to Nojrul, this time to his brusque personality.

Within this stock of candidates, the reputations of Ayan Darogah and the BNP candidate were marred by excess, albeit of vastly different natures, whereas the reputations of Nojrul Chairman and Hatim Master were secure. They were both taken to be good men who, although likely to peddle influence and siphon off from government projects once they came to power, would not hassle local businesses or expropriate welfare meant for the poor, such as rations for the elderly and the disabled. Instead, they were being compared on the basis of how they were raising funds and how much they were spending and their situational capacities, with Hatim's residence on the mainland being to his benefit in this instance.

If one were to venture an initial statement on the first field of influences on chaura voters, they would include the influence exerted by reputable leaders, their patrons, and to some extent party affiliation (after all, both Nojrul Chairman and Hatim Master were Awami Leaguers even if they were not the party's official candidates). One would also say that financial enticements and moral persuasion were readily visible, and physical coercion less so, although this does not allow us to completely discount its presence. However, these influences did not predetermine election outcomes. Everyone, including Hatim Master, was shocked when he won. When I met him in the Nauhata Bazaar on the afternoon of election day, he was giddily splashing *alta* (vermillion) paint on the assembled cheering crowds while being festooned with garlands of taka notes. He kept shaking my hand (a very frowned-upon gesture toward a female but one that seemed to be accepted by the crowds on this occasion) and shaking his head in disbelief, thanking *me* for praying for his success as though only a completely external and unusual presence could account for his good luck. Despite his incredulity, he won by a sizable margin of 452 votes over Nojrul, with 2,965 votes, compared with Nojrul at 2,513 and Ayan with 2,272. If the numbers indicate anything, they suggest that Ayan Darogah might eventually outlive his past actions and bad reputation.

It was the figure of the *jasus*, or spy, who best crystallized the indecipherability of people's voting decisions and election outcomes and alerts us that the field of influences I have sketched here is as yet incomplete. I only met one spy when Shohidul called him out as he tried to hurry past us. He turned out to be none other than Qadir, introduced earlier as someone who dealt in land, a *dalal* of sorts, a middleman in polite language or more disparagingly a tout.[4] He was a wiry character who usually met up with his friends and allies in the upstart bazaar begun by the village of Phulhara after its leading doyen, Yusuf Member, got into an altercation with Shukkur Member of the village of Chaluhara, who resided in Dokhin Teguri and whose "rapport" or directional tendency was toward the more established Nauhata Bazaar where I hung out (see chapter 1). Qadir was usually to be seen on his haunches or with his hand around someone's shoulder, pulling hard on his cigarette while speaking in an intense, secretive manner. On the occasion of the elections he was busy being a spy on behalf of a BNP candidate who did not dare come into AL-dominated areas but who knew there were votes to be had from among a few of the villagers who secretly supported BNP.

Qadir the spy stopped only briefly to urgently whisper that his job was to listen in on the bands of roaming men as they spoke among themselves and to villagers to ferret out any details that could help his candidate. Most important, he had to gauge which way a villager might be planning to vote. It turned out that each of these electioneers carried a list of the names of every voting member of their ward, collected from the local office of the Election Commission. Since the caretaker government had mandated national identity cards with photo identification in 2007, the Election Commission had been able to compile computer-generated voter lists with no or minimal duplication (Nizam Ahmed 2011).

While Qadir the spy was unnerving in his strenuous efforts to know the minds of others, what struck me more forcibly, making me realize that elections in these parts had everything to do with living on moving land, was that some of these men, and quite a few women electioneers as well beginning in 1997, were campaigning for candidates representing villages that had eroded or, in local parlance, were in the river's womb. Their voting population was dispersed far and wide. These electioneers were tracking down everybody from these eroded villages who had chosen to remain on the lists to persuade them of the merit of their candidates and promise them free passage by boat to the voting booths of the relevant wards.

In effect, some chauras were going to vote as if their villages were still intact, as if they were still living there, and as if sometime after the elections they might run into their member representative or chairman in the local market to ply him with cups of tea while putting a plea to him for repairing a country road. For instance, while the wards making up Sthal Union, which saw the dramatic contestation among the five candidates related earlier in the chapter, were all more or less intact, one large village of a ward, which fell within Gorjan Union in which my host village Dokhin Teguri was located, was entirely in the water, and yet people came from all over Bangladesh to vote for it.

What are we to make of this elaborate artifice by the spies, the erosion-displaced villagers, and the politicians? How do we account for these lost villages within the field of influences affecting chaura voting behavior? And what do these acts of voting amount to as gestures? In other words, were they sovereign or not, and if so, how? Before we consider these questions, it is helpful to examine the history of local self-government in South Asia to clarify just what was at stake in local elections.

Local Self-Government and the Village in South Asia

Any description of the administrative structure of government in Bangladesh (as with any modern nation-state) is a quest to master nested categories. While the country has been organized in many different permutations since the establishment of British colonial rule in the mid-nineteenth century, the tendency has been toward further and further division. While the lowest tier of government was the nonelected *gram shorkar* at the level of individual villages, a controversial entity of dubious merit (Ali 2011), the union parishad was effectively the lowest tier elected into power. To arrive at the unit of the union, I find it easier to begin from the top, but it must be remembered that those elected into the lowest tier of government see themselves predominantly in relation to their specific wards or groups of villages and to other union parishads, with the extent of their involvement and influence reaching at most to the level of the upazila, or the subdistrict.

Each of the sixty-four zilas, or districts, currently making up Bangladesh is divided into upazilas. Each upazila is divided into unions. Each union of an upazila is divided into nine wards, with each ward made up of one or more villages. A member, male or female, is elected from each ward. In addition to these nine, an additional three female members, each representing three wards, are elected. Thus twelve members represent a union. In addition to voting for these members, the population of the union also votes for the chairperson, who presides over and represents the union at the upazila level, serving as the electoral college for the chairperson of the upazila. In sum, the union parishad is composed of twelve members and a chairperson.[5]

The earliest initiative taken toward elected self-government derives from Lord Ripon's Resolution on Local Self-Government (1882) that was put into effect in towns in colonial India (Tinker 1968), followed by its extension to villages in 1907. Ripon understood his actions of allowing elections as a means to induct Indians into self-government (Weinstein 2017), while nationalist historians have long pointed out that Ripon's resolution disrupted long-standing town and village management practices through local organizations and figures such as the panchayat, *faujdars, muqaddams, mutawallis, muhtasibs,* and *patwaris* now replaced by local bodies controlled from above, with limited jurisdictions and mandates (Tinker 1968; Gilbert 1990).

While the stimulus for self-government might derive from Sir Ripon in the late nineteenth century, the actual structure of administration in Bangladesh draws directly from the Basic Democracies experiment of the 1950s in postcolonial Pakistan undertaken by the then military leader Mohammed Ayub Khan. Along with his colonial predecessors, Ayub Khan felt that the population needed a more staggered introduction to democracy (M. A. Khan 1967; Karim 1990; Siddiqui 2005). He conceived of a structure by which each layer of administration voted for the one immediately above it, with each layer serving as an electoral college for the one above it. He expected that a point would come when voters would be sufficiently educated on the democratic process to elect leaders at the national level. While at present the general populace in Bangladesh can vote in both local and national elections, this practice of staggered elections or one of having a mixed government, made up of elected and nominated members, is still ongoing. For instance, as described earlier, it is the union chairpersons who constitute the electoral college for the upazila chairpersons, and the upazila chairpersons for the zila administrator. Although the union parishad is an elected body, these elections are accompanied by the creation of a gram shorkar, which is a governing body at the level of individual villages made up of volunteers selected by members of the union parishad.

It is interesting to compare the tasks incumbent on the gram shorkar and union parishad from the colonial period to the present to explore the expansion of responsibilities. The short list of duties of local government in the colonial period placed greatest emphasis on taxation of the local population to fund *chaukidars*, or night watchmen (Siddiqui 2005). It revealed that the maintenance of law and order, one that was self-financing, was the single most important interest of the colonial government in encouraging self-government. This was an early critique extended to colonial-sponsored local self-government by Indian nationalists (Tinker 1968). One sees a distinct expansion of the list of duties of local government in the Pakistan period into the Bangladesh period. Starting in the 1950s, members of the parishad were exhorted to undertake regular surveys of their constituencies to determine their needs and utilize moneys from the parishad's limited revenue base (taxes on built forms, leases of riverbanks, rents of government space) to provide for these needs (Siddiqui 2005). Later, in the 1980s, as local government became of increasing interest to international donors involved in development projects, such as the United Nations Development Programme and the United States Agency for International Development, the tasks of local government grew to encompass teaching people about the ills of society and encouraging them to undertake microenterprise or civic-minded projects. These fell under the rubric of poverty reduction, human resource development, and the creation of employment opportunities (Westergaard and Alam 1995). The mandatory inclusion of women members within parishads that began in 1997, one female to every three wards, was also part of this general push to mobilize local government for initiating social change (Panday 2008; Begum 2012).

At the present, as funds from its limited revenue base regularly fall short of what is needed to run the local government, the parishad receives funds from the central government earmarked for specific domains of activity (Hasan 1992; Siddiqui 2005; Zafarullah and Huque 2001). These *khats*, as the earmarked funds are called, range from waste disposal, to road building, to provisions for rural electrification. These have been subject to the greatest expropriation and now entail a complex system of accounting, regulation, and audit to ensure that the moneys are not "misused," although of course all that means is that expropriation can no longer be petty or individual but rather involves complicated arrangements reaching to the divisional level of government, two levels up from the local parishad. However, as Willem van Schendel (1981) has shown through his study of peasant society in Bangladesh, the early deprivation of regular funding mechanisms to local government put in place a whole system of expropriation of both government funds and those of the general populace. This expropriation was widely accepted as necessary to supplement low government salaries and to help compensate those who provided political patronage. Furthermore, the entire system of checks and balances as it is set up suggests that the central government is well aware of these modalities of expropriation. Even as its ambitions for local government grow, it must put in place the means by which to check this expropriation so as not to have its ambitions confounded.

The current push toward decentralization by international donor agencies and the central government (Zafarullah and Huque 2001) runs as it were on two contrasting views of the realities on the ground. In the literature on decentralization informed by neoliberal privileging of individualized responsibility, we are told that local government suffers because it does not have sufficient funds of its own due to the enforced reliance on central government that makes parishads hard-pressed to be meaningfully independent and attuned to local needs (Hasan 1992; Westergaard and Alam 1995; Zafarullah and Huque 2001). However, at the local level, as far as I have witnessed in the char and in adjoining areas, there are very few efforts to raise funds, particularly through taxes as these are seen as a throwback to the oppression of the colonial and Pakistan periods (Siddiqui 2005). The reliance on central government being quite absolute, these external funds are seen as necessary to keep local government functioning.

This schematic history of local government from colonial times to the present traces the expansion of the ambition for such government, from its previous limited focus on the maintenance of law and order through locally hired watchmen, to making local government the means by which to produce social change and self-sufficiency within Bangladeshi rural society. It also shows that while the picture of what local government should be waxes and wanes, it is shadowed by another picture, that of reliance on external funds and on expropriation, which is widely felt to be compelled and necessary to keep the government running.

These two pictures deepen our appreciation of the context within which local elections took place in the chars by helping us to understand why voters left out of consideration dishonesty in determining who would make a good leader. However, to understand what more is at stake in these elections, we have to make this history of self-government speak to an older colonial preoccupation with the village in India. One might say that the growth of state-sponsored local self-government was in part intended to undercut preexisting arrangements and spatial forms such as villages to replace them with more controlled local bodies, as was reputed to be the case in Lord Ripon's plan and was explicitly the case in Ayub Khan's Basic Democracies experiment. Nonetheless, every stage of the development of local government relied on colonial accounts of village India for justification, harkening to prior times when villages were considered to be sovereign.

Other than "caste," no other sociological object has been as debated as the village in South Asia. As Jan Breman, Peter Kloos, and Ashwani Saith (1997) write, while the village readily suggests itself as the basic unit for economic administration or political representation, it was a long time in becoming so naturalized. And it was colonial officers, most famously Sir Charles Metcalf, writing in 1830, who characterized villages as autonomous republics that endured in India despite what winds blew over them. These representations were later absorbed and readily proffered by the national leader Mohandas Gandhi ([1909] 1996). Other nationalist leaders such as B. R. Ambedkar viewed villages as centers of caste oppression in which people lived together bound by coercion rather than out of mutual reliance (Jodhka 2002). Village studies conducted in the 1950s, after the end of British colonialism, confirmed the existence of such nodes as villages within the Indian landscape but pointed out that it was unlikely that villages were clearly demarcated, autonomous entities. Instead, they were "always a part of a wider entity, subject to the winds which blew from without" (Srinivas and Shah 1960, 1377).

I flag the village in these colonial and early postcolonial studies to suggest that contemporary discussions on decentralizing government in Bangladesh, even when they do not explicitly reference villages, partake in this long-standing debate on the integrity of the village in South Asia and its presumed impulse toward autonomy and self-sufficiency. Even as policy directives push for the self-sufficiency of local government in Bangladesh, questions remain: How capable are union parishads to be autonomous? And to what extent is such autonomy a myth left over from the colonial fixation on village India?

By taking into account this entwined relationship between the history of local self-government and the evolving representation of the village in South Asia, we are better able to understand why elections to the union parishad would be meaningful for the chauras in the context of Bangladesh, what was at stake for them besides factional interests, individual opportunities for economic advancement, the enjoyment of being wooed by candidates, or the possibility of growing local leadership. Elections were the single most important opportunity to make sure one's village counted, to have it be represented, or to make a representation of it stick. But it wasn't as though elections stopped mattering if there were no physical villages.

In fact, the importance of elections redoubled if physical villages no longer existed, for elections might have been the only way to maintain the fiction of them as political units.

We have also come to realize that a village wasn't the bound and thick social organization that it was imagined to have been. Rather, it was more spread out and crosshatched with other villages and social structures (see Bertocci 2002). And village leadership wasn't as unchanging or authoritative as it might have been (Lewis and Hossain 2008). In other words, we cannot presume at the outset that the chauras would come back for their villages, that they wouldn't just move to other villages along the lines of filiation put out by their originating village and become embedded within other nodes in the network. So, elections being the condition of possibility for the continued existence of an eroded village, it was the chauras' attachment to their villages, the influence of their villages on them, that was still in need of description and analysis.[6]

Partaking in Elections

The inhabitants of Kuwait Para of Boro Gorjan village (due south of Dokhin Teguri) only amassed in this neighborhood in 2008 when the three large villages (Bri Dasuria, Kir Dasuria, and Bumuria) in which they lived eroded and fell away into the water within the space of a few months. They crossed union lines to take refuge here, moving from Shonatoni Union and Umarpur Union to live in Gorjan Union. From their accounts it was as if they had crossed an ocean to come to foreign territory.

The displaced villagers of the three villages announced their exodus and arrival by running their boats aground, throwing themselves on the banks of the river just due west of Boro Gorjan, bewailing their fate, and begging the local leaders for benevolence. A female matbor of Bumuria confessed that when she realized that her village was going to break, she traveled for weeks on passenger boats listening in on conversations, questioning people, and trying to catch wind of any available land where she might put up with the group of villagers for whom she had taken responsibility. She learned that the landowners in Boro Gorjan had fields to spare but that they wouldn't be willing to give them up easily, even to lease them. The only way to force the landowners' hand was to overwhelm them with pity—hence the spectacle on the riverbank.

For all their efforts, the displaced chauras were begrudgingly given land that was supposed to be prized agricultural land, but as the refugees pointed out, it was too perilously close to the water's edge to be that. And instead of the customary twenty-four decimals that, as mentioned in the previous chapter, was the norm for households in these parts, they were given either thirteen or a mere seven decimals, each at much higher yearly khajna rates than was usual. But they seized the meager opportunity promptly, with each family setting up house along a straight line down the bank.

When I first saw the neighborhood, it looked like a refugee settlement, with the land bare except for the shacks in a row determinedly facing away from the river. Within a year or two, agricultural fields and home gardens had taken hold, with the houses crowded by jute plants at the front and lush vegetable gardens along the back, with downy, pendulous catkin grass between the houses and the river's edge. These were people renowned for their green thumbs, and they squeezed what they could get from the land, with entire vegetable plots sized up and their produce bought in advance by men from the mainland who ferried the vegetables to the mainland markets.

The denizens of Kuwait Para also felt as if they had crossed an ocean because it felt to them that they had entered a village the likes of which they had not encountered before. Unlike nearby Dokhin Teguri, where there were several prominent authority figures but none who held total authority over the entire village, or even unlike their own villages, where villagers and their leaders were close-knit, Boro Gorjan was under the control of a powerful family, strong in landholdings, in numbers, and in the monopoly of political positions. The Talukdar family of Boro Gorjan had even gotten one of their daughters-in-law to stand for and win one of the three positions allotted to women in the union parishad, a position now held in proxy by her husband. Bri Dasuria–Kir Dasuria–Bumuria men who were used to working in large mills on the outskirts of Dhaka, the nation's capital, when not working on their own lands were suddenly reduced to the status of day laborers to be called on at a whim by the Talukdar family and be paid less than what they were used to earning, a mere BDT250 per day when they were used to BDT400 or more. That they were still unused to this radical shift in their fortunes was indicated by the frequency with which they first identified themselves as chashi (farmer) or grihosto (householder), and then corrected themselves ruefully to say that they were din mojurs, or day laborers.

While they chafed at the control the Talukdar family attempted to impose, they bore with it because at least they had refuge and were not being pressured to become the voters of this village. All of them had remained voters of their original villages, which sometimes won the occasional widow or the parent of a disabled child a welfare card (bhata), entitling them to rice and other rations in their original union, indicating that entitlements crossed union lines.

The tides changed in 2011 when Gorjan Union was finally going to have its union parishad election and Shahid Talukdar, one of the sons of the reigning family, had decided to run for the position of chairperson. He put pressure on the denizens of the sudden neighborhood to become voters of the ward within which Boro Gorjan was bundled. He even offered to defray the costs of making this change at the local election commission office. His trump card was a promise to settle more inhabitants from the eroded villages, which would mean effectively granting legitimacy to Kuwait Para as a home away from home for those from Bri Dasuria–Kir Dasuria–Bumuria. In at least one place in Gorjan Union, an entire village had resettled and reconstituted itself, and Kuwait Para could be the same.

In conversation with several men of Notun Para, they laid bare the tangle of reasons for their decisions to remain the voters of their prior villages. Roshan Ali told me:

> We have no wish to become the voters of Gorjan Union. The reason is because we are the people [lok] of Shonatoni Union's No. 6 Ward's Khridasharia village. Since our village broke in the river, our villagers have been scattered to many areas where we are now living. Our village name persists. It may take one year or twenty years, eventually new char will form. After we move to the new land [jomi], we will need our members and chairperson. To keep our environment [mahal] intact, to keep our ancestors' [baap dadar] name alive, to farm the char land we have to stay the voter of our past union and we have to be the voter of our past village.

Romjan Mistry continued:

> Bri Dasuria, our village, is in the No. 1 Ward of Umarpur Union. It is now in the middle of the river. The people of Bri Dasuria are living in many different places. If we don't stay voters of No. 1 Ward then it will become extinct [bilupto], then we won't have any members, we won't be able to keep our village unit [mauza] straight, people from other areas will not let us go to our land, they will likely seize it [dokhol kora]. Then if we go to the parishad to ask for justice [bichar], the chairman will not

recognize us. No one will give us any value. More specially, we will lose all our land.

In these words and similar ones repeated by many others, we hear something of the temporality of anticipation, the structuration of the present, and the nature of attachment to villages within chaura thinking. While the overwhelming emphasis seemed to be on protecting one's property, the minor note in the chauras' speech indicated that the name of a village was in itself a precious thing to keep alive, to keep one's existence from going extinct. It was what vectored dispersed bodies, giving one value, and ensuring the continuity of one's home environment and heritage. Without it one could not expect to be recognized, much less get justice.

Thus the inhabitants of Kuwait Para used every wile so as not to become voters of Boro Gorjan but also to prevent giving offense to Shahid Talukdar. They pointed out to him that several of them were very sickly and unlikely to be able to live to vote many times, so why waste money in switching their names from one list to the other? They pointed out that they bore the burden of too many women and young children who needed special provisions in the form of medical centers and schools, and they did not wish to encumber the Talukdar family with their needs. In effect they withdrew all claims on the authorities in the area for tending to their needs, turning instead to provide entirely for themselves, even expressing hostility toward NGOs that now seemed suspiciously generous in giving out microcredit loans as if to entrap people.

Determinedly, Kuwait Para dwellers remained voters of their previous villages, and when the day of the elections came, they piled into rented boats to go to the relevant voting booths located in Chantara. Ultimately their voting efforts did not pay off (although in some respects they did, as I will describe later). They all voted for Latif Chairman, an AL incumbent, but were shocked to learn that a BNP upstart had won. There were immediate challenges as well as claims that some voting slips had been thrown into the toilet in the voting center, but the Election Commission accepted the results and the BNP candidate came to be the chairperson. Suddenly their lifeline to their past villages did not seem as vital as before as they had no standing with the new chairperson.

Meanwhile, the situation of Kuwait Para dwellers in Boro Gorjan had become quite intolerable. Apparently, strange men entered Kuwait Para late at night, casting inappropriate glances at the women, with some going so far as to peep into houses at sleeping forms. These men were known to be associated with the Talukdar family, and the people in Kuwait Para wondered if they were being deliberately and intimidatingly needled to cow them into leaving. And leave they would if they had to. Many had already started looking for other places to move to. But, most important, by remaining the voters of their previous villages, these refugees left open a psychic exit, whether it would be back to their villages or elsewhere.

When I visited in 2014, I noticed that the path behind their houses, between their gardens and the catkin grass close to the river's edge, had been raised to serve as a sort of embankment against the river to the west. Apparently their BNP chairman had masterminded the raising of the path to alleviate the stresses on his constituents living elsewhere. The thinking was rife that now that land was to be seen more months out of the year than previously in the area of the lost villages, he anticipated the return of the villagers and sought to curry favor with them by helping them out from a distance. He had to do more than was usual, being a member of the opposition party and in desperate need of a loyal constituency to help him stay on as chairman. This raised path made those living in Kuwait Para determined to leave, to return if possible to their prior villages.

Through this series of mediations, the line of thinking that had become established was that if the elections could just be sustained, villages would one day return. In other words, if the villagers could sustain elections of their lost villages, they could make their villages return. But this first had to work the other way, meaning there first had to be a sliver of land of the original village to hang the elections on. In what follows I provide a second story of partaking in elections through which we come to appreciate the chauras' tortured relationship to their village, in which one might will one's own extinguishing to ensure the continuity of the village.

In Rihayi Kawliya, a village once located on the easternmost side of the char, the question of whether there was even a sufficient sliver to represent a ward caused the elections to be held up for five months. The village, once a large, prominent place, had been in the throes of eroding and accreting from this side of the river shore to the other side that was part of the mainland for the past ten years. It had participated in the 2003 local elections, but then a substantial chunk of the village went to the river's womb. Romjan Chairman, who by this time had served as chairman for eight years, filed a complaint with the Election Commission saying that since an entire ward (Rihayi Kawliya was a large enough village to constitute a ward by itself) under his jurisdiction was eroded, resulting in a tremendous loss of voters, there could be no fair elections. In so doing he took advantage of ambiguities in the existing electoral laws to keep himself in power beyond the end date of his term.

The Local Government Act of 2009, which is the most thorough legislation to govern local bodies since Bangladesh's inception in 1971, makes the following provision for administrative units and constituencies lost to erosion. Article 17 of the act states that in the instance of river erosion, natural disasters, and so forth, the relevant local body presiding over the villages is considered null and void (batil) and that a local body has to be reconstructed/regenerated (i.e., it has to undergo punorgothon) through reelections. But in instances in which the reconstruction process is held up, such as by petition to the Election Commission, the existing local body is to continue as before, militating against its immediate dissolution as suggested by the first part of the article. In sum, in the instance of lacking a ward over which a local body is to preside, the legislation urges one to continue as before, as if the ward were still in existence, while disallowing fresh elections because the ward is no longer in existence. Clearly, there is an aspect of a double bind in this legal formulation.

As a result of Romjan Chairman's petition to the Election Commission, the election was suspended for Gorjan Union (of all nine wards, including Rihayi Kawliya, Dokhin Teguri, and Boro Gorjan) until further investigation. Finally, in large part because of the considerable sums of money expended by Montu, a chairman hopeful, in a case against Romjan and the Election Commission, the commission decided to hold elections when it came to light that there was indeed a strip of land made up of one household and one mosque that were within the original village boundary of Rihayi Kawliya. This little strip of land was felt to be enough to justify holding elections.

The peculiar thing about Rihayi Kawliya was that almost all of its 250 households had shifted to the mainland, pushing into the district of Tangail. This was where Montu, the chairman hopeful, was based, presiding over some two hundred households from the original village. A mere fifty or so households of the original village were still in or around the little strip left of the original village boundary, located on lands belonging to the neighboring villages of Koruajani and Phulhara. Regardless, if one were not aware that Rihayi Kawliya was by and large now on the other side of the river, the villagers who had remained behind acted as if they were still in the original village. Initially I, too, thought that this was Rihayi Kawliya until I realized that most of the villagers in the char spent a lot of time taking boats to the mainland to shop, meet relatives, and make requests to their local big men. Thus, unlike Kuwait Para, in which one was immediately alerted to the fact that the inhabitants felt themselves to be strangers in a foreign land, those of Rihayi Kawliya acted as if their village was still intact with just a river between its two parts, although they were effectively splayed over two districts, Sirajganj and Tangail.

Although the small strip of land that was needed for the elections to be held was in the char on the Sirajganj side of the village, Montu, the chairman hopeful, wanted to hold the elections on the Tangail side of the village. His ambition was to reestablish Rihayi Kawliya as part of the mainland and not part of a lowly char that kept breaking and reconstituting. The voting booth, a disused madrasa, or religious school, building, which had been pushed deep into Tangail when the village was breaking and where it now lay as a stack of corroding tin, shorn of students and teachers, was suddenly the point of attention. Romjan felt it had to be brought to the original remaining strip in accordance with the Election Commission's verdict. As he explained to the remaining inhabitants of Rihayi Kawliya, if they wanted their village to become whole again, it was to their benefit to have the elections held in the char. It would also well serve the rest of his constituency, who were largely in the char, to have the voting center right on the island rather than have to take boats to Tangail to vote. The voting booth was brought back to the grounds of the mosque and rebuilt for use on the day of elections.

Morium, my close informant from the village, told me that Montu's son came to her house with television reporters, who interviewed her for a widely broadcast news show. She narrated:

The journalists came. Montu Chairman's son asked me, "Khala [Aunty], how are you doing?" Then the journalists said, "If she is your Khala then she is also our Khala." They asked me, "Which is your house?" I related that my house was broken by the river and that I had moved it to the western side. "What did you get when your house broke?" "I didn't get anything," [I said]. "I paid a thousand takas in bribe for a VGD card. I went to Shibir [mainland] to have my picture taken. I did sifarish [request intercession] from a lot of people but I got nothing...." Then the journalists asked, "What is the name of this village?" I said, "This is Rihayi Kawliya. This is the Rihayi Kawliya mosque. Those of us living here are all from Rihayi Kawliya village." The journalists asked me, "Do all of you want the union parishad elections?" I said forcefully, "Yes, we want elections. For ten years they have been eating. Aren't there other human beings [manush] [who also deserve to eat]? I want elections." They showed my words in the television. Montu Chairman, Habib Neta [leader], and others said approvingly to me, "We need more chapa folks like you [those with surprising hidden depths?]." Many in the mainland said that woman from the char spoke very well.

Montu, the chairman hopeful, offered the Rihayi Kawliya villagers in the char BDT500 per person for moving the booth piece by piece back to the mainland. It would have been a minor victory for the poorer sections of this village, who had been forced to move farther into the char rather than make the journey to the mainland with their co-villagers at the time of the dissolution of Rihayi Kawliya, to have the voting booth stay on their side. But they took up Montu's cause and helped him surreptitiously move the voting booth back to the mainland, although Romjan, the incumbent, managed to bring the voting booth back again to the char.

Elections were held. Romjan lost, and Montu won but by only a tiny lead, which he likely could have increased had the voting booth been on his side of the river. Everyone agreed that it was important that the elections took place at all so as to uphold the logic of elections sustaining and enabling the return of villages. In the end the remaining Rihayi Kawliya villagers in the char partook in an election that was to enable their village to be elsewhere than where they were, but at least it ensured the continuity of their village.

The Virtual Nature of Villages

It is usually presented as commonsensical that people in largely rural Bangladesh come to vote in their villages. It is taken as such in scholarship, as is the integrity of their villages. However, I hope to have disturbed this commonsensicality by exploring how local elections had come to be related to the village belatedly in history. I also hope to have disturbed the presumed wholeness, integrity, and homogeneity of villages. To draw out the stakes of chauras' attachment to their eroded villages as expressed through their electoral practices, here I return to their own words. This is by way of attending to the earlier deferred questions: What type of influence did lost villages exert on chauras? And how do we understand the meaningfulness for char dwellers of the act of voting for villages that no longer existed?

In the quoted passages provided earlier in the chapter, we learn readily that chauras voted for their lost villages to keep the village intact as a political unit. They did so to protect their long-term interests, specifically their claims on property and farming rights. But they also did so because by some logic that had been inscribed and reinscribed over time, they felt that by undertaking elections, by congregating, they were helping the village to become whole again. Elections acquired a performative dimension. The village was performed in advance of its material reconstitution to aid it in recovering its material form.

The election also kept viable a more existential (as in asserting existence) quality of the village. Shohidul's own village of Bishtipur broke in 1988. Prior to the 2008 national elections, the military-sponsored caretaker government put into effect a nationwide electronic identity card system with the singular intention of reducing voter fraud. Given the fear of military reprisal, no one from Bishtipur returned for its elections in 2011, and without elections, this village was effectively dead, the ward lines redrawn without it. Shohidul pointed out villagers from the now dead village in various parts of the char or on the Tangail side of the mainland, showing how they enacted the death of their village by barely acknowledging one another. Thus the performance of elections was also undertaken in the face of the possible extinction of the village and of villagers in the eyes of others.

This threat of bilop, or extinction, did not exist only in the case of eroded villages. As A. K. M. Aminul Islam (1974), among many others writing on the Bangladesh village, notes, continuing M. N. Srinivas and A. M. Shah's perspectives from the 1950s, it is rarely the case that a village suggests itself through clear spatial boundaries or social structure. Rather, villages typically consist of dispersed households separated by vast fields. They tend to be transected by administrative lines, party affiliations, land and gusti considerations, spatial distances and proximities, and shomaj association. Continuing the work of John Thorp (1978, 1982), Shapan Adnan (1997) has emphasized shomaj as the most important organizing principle of Bangladeshi villages and the one that illuminates the extent to which the fear of extinction haunts relations of Bangladeshis to their villages. The principle is hard to secure, as shomaj was everything and nothing, as a chaura once remarked. It was the group associated with a specific authoritative person or persons congregating in a particular mosque in a village. Consequently, there were at least as many shomaj associations as there were mosques in a village. The shomaj ensured that the ritual calendar was upheld, one of the most important days being Qurbani Eid, during which cows and goats were sacrificed and their meat distributed among the shomaj members and their hides sold to raise money for the mosque. This was one time during the year when the poorer members of the shomaj could expect to eat meat.

Shomaj was subtly much more. Although the feeling of being a part of a shomaj arose on only a few occasions, without shomaj one was effectively a pariah. In the film Titash Ekti Nodir Naam (Titash is the Name of a River), the question of Rajaji, the widow with a child, as to her future in a village rested on whether someone would take her in their shomaj or else she would be cast out of the village (N. Khan 2015). In an occasion from my fieldwork, Sujaat of Rihayi Kawliya, a very romantic figure, once told me that he had left the village because he felt stifled by it and moved his household and family to be a part of the nearby Phulhara village. However, he said in a chagrined manner, Phulhara threw him out for reasons I can only guess given his amatory proclivities. He was forced to return to Rihayi Kawliya, which was duty bound to accept him because he was of the village, meaning he was born and raised there, as were his forefathers. When I asked why he had to return to the village even though he could have gone anywhere, he summarized, Without shomaj life is not possible.

It is noteworthy that the fear of extinction does not arise in the face of an indifferent or cruel river but rather in the face of the state and of society, which was felt to be overwhelmingly composed of those who judge, lie in wait to seize one's belongings, or likely forget one given even the slightest chance. A commonly used phrase to describe the indifference of relatives to one's plight was They don't even turn to look at me. Thus, the influence that villages exerted on voters was that born of anxiety over the possibility of one being extinguished within the register of the social. Elections helped avoid social death.

On the day of the elections, people walked for miles to the riverside, stood under the open sun or in the lashing rain, squeezed themselves into boats that were almost level with the water surface under the weight of humans and cargo, unfurled their crushed selves at another shore, walked miles to a voting booth, then stood in long lines without a break to cast their votes, finally pressing thumbprints next to their symbol of choice before beginning their return journey. As I visited voting centers on election day, Kohinoor, a friend of mine from the village of Borongail but who often came to Dokhin Teguri to help me, greeted me from within a long line at one such center, saying *We have been here before.* I asked her what she meant, thinking she was referring to her participation in prior elections. She cheerfully contradicted me, explaining with nods of encouragement and embellishments from the women standing in line with her that this laborious task of coming to vote mimicked the muscular exertions and physical pathways that she took with her family when her village broke or when it was overwhelmed by waters in past years. As she said, *Then too I had to walk and stand in lines to receive help, food, and cloth from government officials.* The manner in which she delivered up this parallel suggested that this would happen again and that she would keep walking to re-form lines.

This mode of rendering the experience of erosion, floods, and even elections was not one of resignation but a transmuting of what may be perceived as an event external to an activity in which one was engaged. Although the chaura voters were steeped in a field of influences that made their act of voting an anxious necessity, there was still a sovereignty of sorts expressed through this act insofar as they engaged in it actively. This sovereignty was not in the effort of transcending the landscape, of imposing their will on it, or acting autonomously as free agents, but of forging a parallelism between elections and the river's activities. Doreen Massey (2006) writes of the need to see the obdurate elements of landscapes as events to draw attention to their histories and their ongoing presence in time. Extending her thought to the char-river landscape, it is important to consider that elections both perform villages *and* effect events in the landscape—or, put another way, they are events within nature. Through elections, one's ongoing efforts on behalf of one's village become a way for chauras to insert themselves back into a milieu from which they had been excised.

Embedding within the Jamuna River Ecosystem

Earlier I argued that we think of elections as a means of embedding chauras and their villages within the Jamuna. In other words, the electioneering and the arrangements and movements undertaken for voting proliferate human activities across the chars so as to insinuate them in the landscape, to make them a part of the memory of the landscape so that the landscape might materially yield what is immanent in it. How do we ethnographically render this mutual inflection, of rivers displacing villages and of elections inserting villages and people back into the riparian landscape?

As part of my research I collected kinship-related information on several key informants. This involved asking them about their earliest ancestors they could remember all the way to their youngest relative living today. The idea behind doing these kinship charts was to see how far back people could recall their ancestry and how spread out their kinship network was at present. The charts also allowed me to see how many members of existing families had migrated out of the area.

What became readily apparent from these kinship charts was that family names and memories only went as far back as one's great-grandparents. In addition, almost all the great-grandparents and grandparents were born and raised on land that was considered permanent, or qayem. However, by the householders' generation, almost all of them had experienced the breakup of qayem land and the start of the perpetual movement from location to location. At an initial point they still chose to move to locations considered qayem. Their steady impoverishment was tracked by the fact that at some point they could no longer afford to live on qayem land and thus returned to live on new chars in the river. Their hope for moving to these new chars was that once their ancestral qayem land returned as chars, they would be close by and able to reoccupy them.

These charts helped me to understand ethnographically how elections embedded villages within the riverine milieus. While undertaking the kinship charts of Morium from Rihayi Kawliya I realized that her stories and memories of her vast family and those of her deceased husband had an enlivened quality to them. For one, she could rattle off quite a few names of grandchildren and even of some great-grandchildren of her family, particularly those on the side of her father's family, including those she had not seen in years or in some cases had never met. She explained that she was caught up on everyone's lives on account of the elections. She lived right by the *ghat*, or bank, where boats docked, carrying people from the mainland to the char. As such her house was in the path of those who came to the island to campaign and canvass. According to election laws, candidates have a mere six weeks to campaign before the elections, which produces a frenzy of activity of postering, leafleting, and visits by candidates leading up to election day. For example, given Montu Chairman's interest in wooing and winning the chars, during the elections a blizzard of men and women had come to the chars from Tangail on his behalf.

Many of them walked past Morium's house. She stood speaking to them as they passed, offering water or a seat to rest on, pulling those she recognized as her kinfolk into her humble home. While there were many such meetings, I present two in particular that allow me to speculate on how electioneering intercalated with the landscape. Among the women who walked by were two young women, Najma and Nasima, the great-grandchildren of her grandfather's brother, one newly married and one who had just passed her secondary exams. They had been brought to the chars by Hossain, Najma's new husband, as someone from the distant reaches of Morium's family was running for the position of union parishad member of Rihayi Kawliya. Morium grabbed them, begging them to stay a day or two with her, so they passed the night with her at her home. Morium described in detail the pleasure of talking with them until late at night, hearing their stories of the other members of her father's family. In the morning they continued on their way.

In the second instance, Mustafa, Najma and Nasima's brother, came with a band of people to drum up support for the candidates from across the water. However, Morium said worriedly, he didn't seem much interested in speaking to people about the candidates. He just walked around Rihayi Kawliya to find out how much land had come up in the recent monsoons. It appeared that he was involved in a deal in which he had leased his land to someone residing in Rihayi Kawliya. Morium speculated that Mustafa suspected he was being defrauded by the person who had leased his land or that Mustafa was considering the viability of his land for a potential return to the chars in the future.

The elections came to an end, and all the people who were busily out and about withdrew. Morium was left with an enlivened memory. She now held almost a complete roster of her ties to people, some thickened by contact and care, others winnowed and weak. She knew of their coordinates that enabled a mental map of their spatial relations to one another. And she knew of the hopes and anticipations of those who came through and looked around. Besides the number of people from the other side of the river who came and made pathways during the elections, they also left their mark on the landscape through Morium's mediation. I would claim that these were further means of embedding oneself within the Jamuna, with Morium as their stylus on the landscape.

On a return visit to the chars in 2014–15 in the midst of the chaos of the national elections and violence in the entire country, I walked around as usual to see the changes since my last visit. Although I was well aware of the constancy of change in these parts, I still had to moderate my entrenched expectations of continuity so as not to be overwhelmed by the radical nature of change. Thus, for instance, my heart sank when I realized that ten out of the fifty or so households associated with Rihayi Kawliya had moved as the river's bank they were living on had almost entirely fallen off, leaving a sharp precipice. I had to scramble over thin, protruding ledges to move between the remaining households.

In the midst of this I was shown two faces of the local government I had seen elected in 2011. In Rihayi Kawliya, or what remained of it, I observed nothing but scorn for the members and Montu Chairman, who had been voted in. For example, a woman whose house was one of the remaining households I visited told me that before the elections Montu Chairman gave her his telephone number and said that whenever she would call him, he would answer, call her *ma* (mother), and listen to her concerns with promises to help. She tried his number as I watched to show me how he refused to take her call. She spat to the side, calling him a motherfucker. It was as if Rihayi Kawliya, at the brink of eroding again, had fallen into the politician's blind spot. If there were anything to be done, it would have to be done on the other side of erosion, not before it. Meanwhile, these people waited to have their households wash away while a few with some means tried to make alternate arrangements.

The members of Sthal Union (just north of Dokhin Teguri) who had voted in Hatim Master had been busy. As I walked about I was shown how people had started to return to live in Bayisbari village, which was only freshly up when I was there in 2011 and only lightly populated when I returned in 2012–13. With more people returning, there was a need for houses to be upraised, tube wells to be dug, and latrines to be established. While people generally had to bear the entire costs of these projects, leading to houses not being fully raised to the recommended level, the area around the tube well not being properly cemented, or the existence of informal toilet arrangements, the union members of Sthal had seen fit to make some provisions for the returnees. So, there were now intact households in the midst of what had been just fields a year earlier.

On further inquiry, I found that no fewer than thirteen projects had been undertaken in the conjoined areas of Nauhata village, Bayisbari village, and Chaluhara village that made up the Sthal Union ward. These thirteen projects were all in some way engineering projects, including road building, latrine building, upbuilding of household plinths to make them flood-proof, flood shelter construction, the creation of prayer spaces for Eid prayers, the repair of offices and mosques, and so on. They were at one level about improving the infrastructure of these villages, to improve connectivity, and to ensure protection against disasters. But they were also about the cutting, dredging, and transport of earth. They constituted interventions into the landscape. The member of the union parishad with whom I spoke admitted that in every instance anywhere from 20 to 40 percent (his own calculations) of the projected work had been completed (it being unheard of that the full resources would have been applied to carry out the work). Insofar as even this much had been done, it meant that there had been considerable tweaking, shaping, and pummeling of the landscape to render it more labile for human use, which may be considered another kind of embedding within the milieu produced by elections.

This intervention into the landscape was scaled up to the subdistrict level, in which there was considerable governmental and commercial activity premised on providing defense from the river and protection from natural disasters. But this activity hadn't been enough. The town of Chauhali, the seat of the subdistrict administration one up from the union parishad administration of Gorjan Union, was on the mainland along the eastern side of the char (on the side of Tangail district where Rihayi Kawliya was currently reassembling). Over the four years I had been going to this area, I had seen this town change from being a relatively flourishing one with a long stretch of markets along the riverfront and tall government buildings in the background to one that was almost entirely in the bosom of the river, with only a few shops along the banks and all government offices in temporary quarters farther inland. The first defense that I saw put down consisted of cement slabs that were stacked along the riverbank to reinforce its slope, similar to the Monakosha Badh on the mainland along the western side of the char. However, as the waters gouged the slab, sandbags replaced them. As the sandbags started slipping off, cement blocks were thrown pell-mell into the river as if to slap whatever hands were creeping up on the town.

When I asked why this area hadn't been declared a disaster zone, as had the hard point (the rock dike used to stabilize the river bank) in Sirajganj

Town, which had similar problems, and why the military wasn't out preventing its dissolution given its significance as the seat of subdistrict government, I was told that there was no will to save it, that the political and commercial forces would rather keep the situation as it was because it effectively sustained the commerce and industry of protection without actually providing it.

At present the cement blocks were being provided by the Water Development Board, the government institution reputed to be as corrupt as the Land Ministry and the Forestry Department, the sandbags by those in the AL party machinery, and the sand by influential private contractors. The connections between this industry and electoral results were clear. The minister Latif Biswas's wife had become the chairperson of Chauhali investing in commercial activities, including the dredging of the river for sand for construction and the making of cement blocks. But it was clear to all that neither Biswas nor his wife was motivated to do much more than profit from the endeavor, as Biswas was more identified with the western side of the char and wanted his legacy to be making roads rather than embankments to service the commercial interest for which Belkuchi, his prosperous village, was known. Romjan, now the ex-chairman, had once been in cahoots with them with a contract from the Water Development Board to make blocks. His replacement by Montu Chairman could have indicated a change in motivation as Montu was more identified with the mainland alongside Chauhali on the eastern side of the char. However, the change in the AL's subdistrict party formation had left Montu without any influence within the subdistrict administration other than as a chairman of a union. Consequently, Chauhali was awash in exhaustion and a similar sense of inevitability that I perceived in the remaining households of Rihayi Kawliya across the water. While previously it had seemed that the river was hell-bent on coming down this chute between the char and the mainland with destructive force, it now became a bit clearer that the river was being allowed to be so destructive.

Shariat Kha (Khan) of Rihayi Kawliya, a bear of a man who was both a boatman with a large transport boat of his own and a farmer on lands that he leased within Phulhara village, provided me with the most vivid description of the Jamuna River system as I have come to understand it with the mutual imbrication of the social and landscape in these parts. In recounting the damage being sustained by Chauhali, he mentioned 1996, the year the Jamuna Bridge was coming up, effectively connecting northern Bangladesh with the rest of the country. After the bridge was erected, the land southeast of it started to erode, beginning with Sholosho Jangali. Then went Mishtighati, Sthal Char, Moheshpur, Boyalkandi, and Barbala. Rihayi Kawliya fell too as part of the movement of the river and the eastward shift of its currents. Chauhali's fate was apparently already clear at that time. Latif Biswas, Romjan, and Hatim Master were all playing their parts in hastening its destiny. In fact, Shariat Kha said, shrugging, he had been waiting a long time for Chauhali's destruction because only with its dissolution could Rihayi Kawliya finally come up. So while parts of the village had fallen into the river, to the dismay of its occupants, large stretches had come up that belonged to Shariat Kha and other hopefuls like him. At the same time, it was the dissolution of Chauhali that was also leading to the reconstitution of Bri Dasuria–Kir Dasuria–Bumuria farther south, giving hope to those stranded in Kuwait Para of an exit from Boro Gorjan and the oppressive Talukdar family in the near future.

None of these courses and events were causative in any predictable sense given the capricious nature of any braided river, particularly the Jamuna, which is considered a machinery of complexity. An old chaura once told me that the Jamuna was more mysterious than his wife. Yet it seemed to me while speaking with Shariat Kha that by partaking in elections, chauras had been thinking of the Jamuna system and stitching themselves back in to the system by means of the elections.

Conclusion

This chapter has attempted to show how an apprehension and reconstruction of the char-river relationship within the Jamuna ecosystem might be at work within chaura participation in elections. As I have claimed earlier, the chauras lived conditional lives in which they thought and tried to plan for different courses for every near and far horizon to the extent that their resources would allow. I hope I have already shown that the differential built into every course was not insignificant, a product of the economic and political precariousness of the chauras, the physical dynamism of their surround, and the flexibilities built into their social institutions, such as kinship and property relations. From one perspective, elections were simply one more technology by which to ensure the continuation of the scaffolding of their lives, in this instance the political unit of the village in the absence of land. However, from another perspective, the joyfulness and the curiosity for details with which elections were engaged suggested a further emotional and imaginative engagement with them. The canvassing for elections, the enlivening of relations through them, and the traveling to voting booths were always accompanied by the memory of chaura dispersal at past events of erosion and the idea that the elections constituted a return to these previous moments, an imagined drawing in and rebinding of all the dispersed particles of human existence. The ongoing speculation over which politician would do what after the elections and what chain of actions this might put into play with each possible action suggested a further gloss on conditional lives. The chauras' multiple uses of the elections to imagine themselves drawing in and rebinding provide a perspective on the Jamuna River as a system, making the chauras a part and the elections a process within it.

The elections help us to further understand how an entirely human product can be a part of nature. Previously, we saw how the river worked upon the chauras as it would sediment, taking them along particular courses where they organized themselves according to the ecological memory of the ecosystem. I posited that chaura self-opacity and the distinct morality that prevailed during these experiences of erosion was, as the trace of nature, not an external determinant of chaura lives but in their unconscious as an impersonal force. In this chapter I venture that the chauras participate in and narrativize elections as a way to reproduce the Jamuna system in the mind's eye. In doing so they reinsert themselves into the system. I take this as an instance of the power of their imagination to attempt to comprehend *and* participate in the workings of nature in and around them.

Awami League posters speak of the
reach of political parties into char
spaces. Photo by author.

People on their way to vote. This
boat was mainly reserved for women
voters. Photo by author.

A tired but triumphant Hatim Master
was adorned with garlands of money
and showered with colored water
to celebrate his election win. Photo
by author.

Chauhali shodor (town) was propped up with a makeshift embankment of sandbags. Photo by author.

Decay of

the River and of

Memory

Those Who Do Not Return

THUS FAR, WE HAVE LEAPT MIDSTREAM into chaura life in the Jamuna River with fights over land dissolving on the occasion of the erosion of char lands, with those people displaced by floods and erosion streaming across existing lands, and with chauras attending to their absent villages through elections in the hopes of their future return. Interspersed among these trajectories were scenes of the return of people once land reemerged. Mostly the poor returned at the earliest stages of char formation to eke out an existence among the kashbon, or catkin grass, by growing cheena, or peanuts, that were easily cultivated in sandy land, grazing the cattle of those who lived on the mainland, and fishing. Through their settlement and cultivation of more kashbon, komli, another plant that readily takes to chars, and various grass varieties, some of which return on their own with the char lands, such as the napier grass, or elephant grass, introduced to the area a decade or so ago but now endemic, the sandy soil acquires organic matter that aids in its binding and becoming available for further cultivation.

In the next stage in the life of a new char, which might have occurred simultaneously with the earlier stages, people farmed collectively, regardless of the ownership of the land, with crop yields divided according to the contribution of capital and labor. It was at a somewhat later stage, two to three years after the initial emergence of the char, when the char appeared to be durable, that the owners of the land returned or asserted their ownership rights through others, thus initiating a rough division of the land according to the settlement maps produced at the time of the British based on the cadastral survey of the early twentieth century (see chapter 1). As the char endured, more of its original inhabitants returned and insisted on reclaiming their land. At this point, the village matbors might have hired an amin or someone with knowledge of how to measure land whose task would be to undertake a more rigorous division of property, according to the settlement maps dating back to the Pakistan period, the Settlement Attestation of the 1950s. In a few instances of char land that had actually been settled and recorded in the Bangladesh Survey begun in the 1970s, still ongoing in the 2010s, the amin might follow the dictates of the Bangladesh Survey rather than those of the Settlement Attestation, but such a situation was rare in these parts as little of this land was above water in the early decades of Bangladesh. Skirmishes over village boundaries occurred much later, after the consolidation of personal property holdings and the securing of a village political structure. These tensions over boundaries were an important means by which villages, effectively returned from the dead, reasserted their presence and reclaimed their previous identities and their relations to adjoining villages.

In effect, each char undertook a repetition of history, returning first to a communal mode of using land, then carrying out through successive stages the various land settlements and settlement of village boundaries from the nineteenth century to the middle of the twentieth century only to ultimately reach a stasis, joining others in hoping that the government would one day conduct a fresh survey so that all the transactions that had taken place since the 1950s and had been duly recorded on official-looking stamp papers in various local land offices might make their way onto the new settlement maps of Bangladesh. One might say that time in the chars was out of joint with that in the rest of the nation-state, with chars returned to or stalled at an earlier historical stage, while time in the rest of Bangladesh moved forward.

Although neighbors and relatives provided shelter in the immediate aftermath of displacement, such hospitality was not indefinite, and thus those without land had to make other arrangements after a period of taking shelter. They might go to newly emerged chars to live communally, but once the chars became more permanent and informal land settlements took place, they had to decide what to do. If they chose to stay, they had to enter into a verbal rental contract in the form of khajna (annual rent) or kot (a lease for several years in which the lessee paid a large amount of money up front with the expectation that some percentage of the money would be returned when the lessee returned the land to the owner within a set period of time or earlier). Or they could enter into borga, or sharecropping arrangements, which are more traditional for this area (van Schendel 1981; Westergaard 1985). They made these arrangements with the owners of the land or their proxies, if the owners were living elsewhere. If they could not afford any of these options, they squatted on the lands of those who hadn't returned. This immediately raises questions of who didn't return after an eroded village was reconsolidated, and what became of them in terms of their actual lives and in chaura memory.

In asking around about those who did not return, I received a plethora of interesting responses. Clearly, those who had money, education, and possibly government jobs were not returning, as the chaura way of life was considered very unstable. If one had the means, one opted out of it entirely by relocating to some other part of Bangladesh, such as farther north to Rajshahi; to nearby urban centers in Tangail, Bogra, and Pabna districts; or to Dhaka, the capital city, which was a major draw. However, even those who had permanently moved away did not give up all holds on their property, maintaining proxies in the shape of younger brothers, sisters' children, and so forth (see chapter 1). Their reasons for doing so were varied. Some did this out of an enduring tie to their ancestral lands, others because it was an investment of sorts and land aided in maintaining kin and patron relations that were so important to shoring up one's position in a country such as Bangladesh (see chapters 1 and 3).

Jolahas, or weavers—specifically those who owned weaving equipment— also tended not to return if their villages were reconstituted on island chars. The reason was clear. They needed to be close to markets for orders and supplies and being on islands put them at a huge disadvantage. Many more of them preferred to sell the land, in part because they did not have the relatively deeper financial pockets of those who had the capacity to maintain landholdings that were only intermittently above water and intermittently productive.

It also became clear that it wasn't only humans who were not returning. A few people commented that dolphins and crocodiles no longer returned, and that their disappearance had made the Jamuna a much milder river than previously. They didn't mind this disappearance as much as they minded the loss of vultures, on which they had once counted to pick clean the carcasses of dead animals that now had to be dragged to the river's edge to be thrown into the waters. They explained that the vultures had likely been poisoned by the chemicals that now ran through the bodies of animals, such as cows and goats, and that by extension human bodies must also be poisoned. Finally, ghosts had stopped returning. Previously, as soon as the land came back, one or two ghosts were invariably sighted walking around, or rather floating as they seemed to hover over the ground, weeping in joy that their beloved place of origin had returned. But now they were rarely to be seen or heard calling to unsuspecting passersby from the kashbon as they had done in the past. While time might be out of joint in the chars, char dwellers were nonetheless implicated in the global state of decline in biodiversity, rise in chemical pollution, and secular disenchantment.

There was one group of people whose absence from the villages was also remarked on, although more rarely. *There used to be many Hindus living among us in the villages*, an older chaura mused. Wajib bhai, a middle-aged, lively chaura happy to burst into theatrical songs at any moment to show off his baritone voice, repeatedly said, *When the Hindus were here, we had thirteen festivals in twelve months.* This was a popular way of saying that there were more rituals than months of the year. In another part of the char, Nuru Ghosh, a retired amin, recalled the large, stately buildings that served as the homes of Hindu zamindars. Many abandoned these buildings as they left for India. The buildings remained in place, breaking into bits and pieces until they fell crashing into the waters.

Rich, poor, zamindars, *chamars* (those who worked with animal hides), Hindus were here in plentiful numbers as recorded in British imperial gazetteers and local histories, but when young chauras told me they didn't believe there were ever any poor Hindus in the area, because they had only heard tales of the rapaciousness of the archetypal Hindu zamindar, I realized there had been a shift, a loss of memory of a shared existence, however tense or antagonistic that existence might have been. It wasn't that memory had become denuded because of the decreasing chances of real-life interactions with Hindus given their scarcity on the ground. After all, a young chaura could recount in wonderful detail the experience of swimming with dolphins, when in actuality their grandparents were the last in their families to have swum with dolphins as young children. There was a purposive forgetting in the chars that was well aligned with the national trend toward differentiating from India and emphasizing the Muslim Bengaliness of Bangladesh that had increasingly been in evidence almost since the foundation of the country in 1971 (Menski and Rahman 1988; Shehabuddin 2008; A. Chowdhury 2009). There was also the chaura interest in the Enemy Property Act of 1968, with which the Pakistani authorities once seized the properties of anyone suspected of being an enemy of the state largely through their suspected association with India. The Bangladesh authorities attempted to ameliorate this draconian act with the Vested Property Act of 1974, although this act still left property vulnerable to seizure by the state if any member of the family migrated to India (Guhathakurta 2012). The land so seized acquired the status of khas, or state property, and was availed by many ordinary people, including the chauras, through informal and illegal processes discussed in chapter 1. Most recently Hindus have emerged as the minority whose very presence indexed India, an intolerable foreignness within Bangladesh's sovereign territory, making them the object of state-sponsored and non-state-sponsored threats and intimidation at political flash points.

There was something a little different in the chars than on the mainland, a kind of memory and forgetting forged in relation to nature to create what Simon Harrison (2004) has called "memorious and forgetful landscapes." Harrison shows how the Iatmul and Manambu riverine people, shorthanded as Sepik folk and based in the meander belt of the middle Sepik River in Papua New Guinea, evince as much forgetting as memorialization of their past. This is in part because of their ambivalent relationship to their past, which may have been conflictful and thus not useful to recount in the present, and their conception of what knowledge is worth remembering in relation to those in power. Harrison's original contribution is to show how remembering, or rather forgetting, is also aided by the landscape, which is very changeable: "They may also give their landscape a vital role in this purposeful amnesia" (136). This is not to say that people are thereby compelled into forgetfulness by their shifting landscape. The very same landscape that aids forgetting also helps Sepik folk organize past events chronologically in their oral histories. Rather, they elicit the activity of the landscape, particularly the actions of water, to aid in forgetting some things over others.

I have dwelt here on Harrison at some length to show how the Sepik he portrays don't harness the landscape to their use, because such control over the landscape is beyond them, but rather elicit its participation in their projects through allowing its processes to continue unhindered or inflecting it through the creation of small diversions or covering over its activity with camouflage. It is this nature of interrelation between one's surround and one's state of remembering and forgetting that I find between the chauras and the char-river landscape with respect to the Hindus of the area. I am also mindful that in an interreligious milieu, religious others enter into one's subjectivity and interpersonal relations, such that their excision and forgetting linger as poisonous knowledge within (Das 2006, 2014).

In Dokhin Teguri, where Wajib bhai lived, there was a crack in the landscape at a particular point at which everyone swore there once had been a temple dedicated to Kali, a Hindu goddess with a fearsome appearance who was considered a vanquisher of evil forces. It is noteworthy that she was not marked as Hindu in the past as she counted both Hindus and Muslims among her devotees, but she was distinctly Hindu in the present (A. Roy 1983; Hoque 1995). Even though this village was on its umpteenth reappearance since the time the temple had been a physical reality and every current resident in this version of the village was Muslim, in every prior and present version of the village this site was considered inauspicious. Children claimed it was haunted and avoided it at all costs. What was further noteworthy about the goddess Kali was that she was Kali, the black-faced one, and Kala, of time in its fullest expression and thereby of nature in its changeable aspect that brought life or death to all things. The return of the village and the return of the memory of the Kali temple with it seemed to me an event outside of historical time, an encounter of human memory with nature's becoming through the mediation of the iconography of an other.

Or, recall Nuru Ghosh's earlier story about the zamindar's building that stayed on after the zamindar had left the country. The zamindar family was none other than the much-feared Sthal Pakrashi bongsho, notorious for their astrological prowess in telling the future. Their physical compound produced fear in those who lived in its environs as they claimed to hear the cries of those imprisoned for their various transgressions in the cellar that was said to have extended three floors into the ground. Nuru Ghosh had also heard of the Hindu women kept cloistered within the compound, and he imagined that they crept about silently, emitting only the tinkling sound of the jewelry that they wore on their ears, necks, and ankles. Even after the Pakrashi family left for India, people claimed to hear plaintive cries from the abandoned compound. Nuru Ghosh believed he could hear the tinkle of the jewelry on certain nights when the breeze blew across the sandy plains. As the building broke bit by bit until it fell crashing into the water when he was thirteen or fourteen years of age, he finally felt released from that oppressive presence that had become more ominous with each passing year. The ruins and their final dissolution by the river were his salvation from a past that had otherwise weighed heavily on his youthful imagination. If previously a temple site kept returning to haunt, in this instance the disappearance of a hated building reduced the weight of the past. Can we consider that in both of these instances there was a way in which that nature was embroiled, even complicit, in human projects, in this case facilitating the action of forgetting?

In continuing the conversation that I have been staging between Schelling and the char dwellers across this book, I turn to Schelling now to see how he might understand not only forgetfulness toward others but also nature's participation in such mental processes. After Schelling was done with thinking about the relationship between mind and matter—with nature as the unconscious within humans and as the link between mind and matter, recast sometimes as thought and self-organization—he turned to considering human freedom, to ask how freedom is even thinkable given the degree to which the human is bound up with nature. In *Philosophical Investigations into the Essence of Human Freedom* he posits that acts of evil are instances of human freedom. Schelling writes: "Freedom is the capacity for good and evil" ([1842] 2007, 23). However, so as to avoid making nature a stand-in for God and allow it to have independent existence (and, undoubtedly, to also prevent the association of evil with God), Schelling suggests nature be understood as a yearning: "The yearning is not the One itself but is after all co-eternal with it" (28). And, farther on, he states, "The understanding is born in the genuine sense from that which is without understanding. Without this preceding darkness, creatures have no reality; darkness is their necessary inheritance" (29).

While this line of philosophizing paints nature as deeply theological in its desire for God, it continues Schelling's commitment to thinking of nature in desubstantialized ways: as force, thought, and now yearning. And his phrase "darkness is their necessary inheritance" encapsulates nature as not just external but within the human unconscious. As such nature begets good, for instance in the shape of chaura morality during the time of erosion (explored in chapter 2), and can move humans to evil.[1] Within this framework, the forgetfulness that char dwellers evince toward Hindus and the baleful effects on Hindu lives that come from this forgetting could likely be understood as human freedom exercised as evil. While a theologized nature, as tied to but still separate from God, would make perfect sense to char dwellers, within their own framework they would reject the designation of their acts of possessing Hindu lands and forgetting their shared lives as evil. The chauras prefer to cast their acts as a necessary injustice that they commit more by force of circumstance than by ill intent. Thus, they help move Schelling's understanding of a theologically charged nature to that of nature marked as much by decay and ruin as by creation and productivity evidenced in the preceding chapters.

This chapter is an exploration of what does not come back, as well as those processes by which rupture and forgetting might happen. I examine historical texts, notably imperial gazetteers and a local history, that show Sirajganj in the period between the 1880s and the early 1920s when the district, at that time a subdivision of Pabna, was still intact against the ravages of the Jamuna River. The description of a mixed sociality stands in sharp contrast to people's attenuated memory in the present and what I call the active absence of Hindus within chaura lives and char landscapes.[2] The term *active absence* refers to the dematerialized presence of Hindus, say, in the form of ghosts and hauntings, of which I have already mentioned a few. The spectral presence of Hindus suggests a wavering realm of mutable relations with them. But active absence also refers to the fact that the absence of Hindus was the condition of possibility for important aspects of chaura lives, notably being able to stay put and getting by with land for living and cultivation. I draw on the perspectives of contemporary Hindus living on the mainland, a few of whom I came to know, as they looked back at the chars they had left behind and spoke of what it meant to be an archaism, consigned to the past yet still very much alive and present in Sirajganj. In the last section of the chapter, I consider a film fictionalizing Bengali Hindu estrangement and out-migration from East Pakistan, describe a visit to the ruins of a famous zamindar house in Tangail alongside the Jamuna, and read a ghost story based on one such ruin to draw out how Muslims and Hindus came to be delinked in these parts and how we might consider the (un)ethical dimensions of Muslim chaura actions of withdrawal, forgetting, and dispossessing in step with nature's decomposing into rot and ruin.

The Past of Sirajganj

There are two ready sources on Sirajganj's past that provide a glimpse of the district, but each with its own distinctive mode of exposition. On one hand we have the 1887 and 1908 gazetteers written by British colonial officers that strive to provide an impartial summary description of Pabna district, of which Sirajganj was once a part. On the other we have a local history of Sirajganj, written by Maulvi Mukhtar Ahmed-Siddiqi in 1916, that mimics colonial efforts at a neutral description, even citing numerous colonial writings, including the gazetteers, but aims to be more comprehensive, incorporating, for instance, details on bridges and schools and providing more local flavor by incorporating folklore, biographies of men of distinction, firsthand accounts, and so forth. This kind of self-conscious effort at marking the distinction of a locality by providing it a history is all the more interesting given the words of Sir William Wilson Hunter, one of the authors of the gazetteer, who writes: "Pabna District is a comparatively modern creation of British rule, and possesses no real history of its own" (1886, 512). Be that as it may, it is beyond the scope of this chapter to delve deeper into the construction of Pabna and Sirajganj as specific localities. I focus instead on the accounts the sources provide of a certain past of Sirajganj—specifically, its mixed interreligious sociality in the late nineteenth to the early twentieth century (1886–1926).

In the 1886 *Imperial Gazetteer of India* written by Sir Hunter we learn about Sirajganj as an entry on the district of Pabna, as Sirajganj was then a subdivision of Pabna. The district of Pabna had been carved out of Rajshahi in 1832, and the subdivision of Sirajganj out of Pabna in 1845, to make administration more efficient. This meant that these were new administrative entities with new centers of local government. At these centers, notably Pabna Town and Sirajganj Town, stranger sociality was the norm with transplants of British colonial officers, Marwari traders, British merchants, and laborers from up-country, with Pabna and Sirajganj serving as an entrepôt for trade in North Bengal. By the time of the 1908 *Imperial Gazetteer* entry on Pabna, however, written by James Sutherland Cotton, Sir Richard Burn, and Sir William Stevenson Meyer, Sirajganj was waning as the center of trade. The jute trade for which Sirajganj had been well known was on the decline, with its jute mills closed after an earthquake and the shifting of the course of the Jamuna River some miles away from Sirajganj, although all the jute from up north was still being brought there to be pressed into bales before being shipped to Calcutta, West Bengal.

Both gazetteers mention the ecological dynamism of the region. The entire area was crisscrossed with waterways with alluvion and diluvium constantly taking place. While Sir Hunter communicates this variability and changeability in broad strokes, Cotton, Burn, and Meyer are more discerning. In their 1908 gazetteer entry we are told that Pabna was similar to Rajshahi farther north, with "silted-up river beds, obstructed drainage, and marshy swamps " (304), whereas Sirajganj had better drainage similar to Bogra, making it a relatively healthier environment for humans. Small topographic differences were seen as determining the favorability of one place over another. Consequently, Sirajganj had the denser population. Just as in 1886, in 1908 Muslims constituted 75 percent of the total population of the district, identifying themselves as Shaikhs, Jolahas, and Kulus. However, the 1908 tally did not mention any Brahmins among the Hindus as in the prior gazetteer. Among the Hindus, Namasudras or Chandals were in the majority, followed by Malos, Kayasths, and Sunris, all considered depressed castes. And agriculture was the predominant activity followed by industry, commerce, and other professions.

The 1908 gazetteer provides little insight into the sociality of the rural hinterlands and is circumspect in its mention of the history of peasant rebellion in the area. It notes that while Pabna was infamous for the agrarian riots of 1873, the riots themselves were insignificant and quickly put out by British forces, but generated discussion and the creation of a raiyat, or peasant charter, enshrined in the Bengal Tenancy Act of 1885. In contrast, Hunter's 1886 account is much more detailed and worth revisiting for the flavor of Muslim and Hindu relations that it provides, with the close identification of the two communities helping to explain the noncommunal nature of the riots. Hunter writes that the Natore Raja's zamindari holdings had to be sold to five new zamindars on account of his insolvency. Recall that the Permanent Settlement of Bengal of 1793 had created the zamindari land system, and the problems with this system were well apparent by the late 1800s. The new zamindars, Hunter notes, were not well liked because of their efforts to raise both rents and taxes. The raiyats went on rent strike, taking up the matter in court and through insurrection. The end of the riots saw the Bengal Tenancy Act, which the raiyats hailed as a success and which Hunter hoped would ameliorate the chronic relationship between the landlord and tenant that had long plagued East Bengal. He notes that the Muslims and Hindus of Pabna were very intermixed: "The Muhammadans of Pabna do not appear to separate themselves strongly from their Hindu neighbors" (514). Further on he writes that "the lower class of Musalmans mix freely with the Hindu lower castes, and it is said many Muhammadans take observance of Hindu religious festivals (pujas) while among the Hindus certain classes honour the festival of the Muharram, impartially with those of their own Durga or Kali" (515). It is clear from his writings that it is the recent past of the rebellion that informs this inquiry into intercommunal relations. However, he warns that conflict between the two communities may yet be forthcoming as a result of the nineteenth-century Faraizi movement, the Islamically inflected peasant movement that was contributing to accentuating religious differences between the two groups.

Ahmed-Siddiqi's Sirajganj Itihas (1916) goes into considerably more detail on the background of the riots, which he calls uprisings, and in its insistence that looting was not the intent of the rioters as ostensibly claimed by some. He recasts the agrarian riots of 1873 as the Tenant Revolt and cites as its inspiration the 1857 Sepoy Rebellion in which the Indians in the British army rebelled against their officers. He also cites the 1859 Indigo Rebellion in which peasants rebelled against the forcible farming of indigo as inspiration for the Tenant Revolt. In other words, while in their entries the colonial officers see the peasant riots as local conflagration against the excesses of the landowning class, Ahmed-Siddiqi sees them as part of a long-standing conflict against British rule.

Ahmed-Siddiqi also provides long profiles on the reigning zamindari families in Sirajganj. His list includes the Sthal Pakrashi lineage, a name that has crossed our narrative several times in this text, specifically in Nuru Ghosh's memories. In Ahmed-Siddiqi's lengthy account we learn that this lineage originated in Horidev Bhattacharjo of the Jessore district, which was farther west. Apparently Horidev was a well-known astrologer whose reputation brought him to the attention of none other than the Natore Raja, of whom Hunter speaks. Horidev made astrological calculations for the raja about his future, and when they came to pass, the raja gave him twelve mauzas, or administrative units in present-day Sirajganj. Horidev went on to live in Sthal, one of the mauzas, and had six sons. His sixth son, Tara Chand, had a son, Shoba Ram, who tired of taking care of the family property and acquired the certificate of being a Brahmin pandit. Consequently, he took the title of Pakrashi while the rest of his extended family continued to bear the name of Bhattacharjo. His branch of the family henceforth came to be referred to as Pakrashi and to be strongly associated with Sthal, the area where they lived, hence the name of Sthal Pakrashi for their lineage.

Ahmed-Siddiqi's account is interesting for several reasons. First, we see how Sthal Pakrashi is associated with the Natore Raja, also mentioned by Hunter, but we get a different story of that relationship, one of royal patronage, whereas Choudhury (2001) informs us that this bongsho was indeed one of the five zamindars that bought the raja's holdings and later incited the peasants to rebellion against them through their heavy-handed taxation. Second, the account is almost entirely focused on the religious commitments of the various members of the bongsho. So, for instance, we are told that Horidev Bhattacharjo, the original ancestor, espoused a preference for the more martial of the gods, erecting statues of Shiva, Narayan, and Ganesh in Sthal upon moving there. Later Shoba Ram, the original Pakrashi, was more persuaded by the love of the mother and erected Kali temples, perhaps one of which keeps returning as a haunted site in the nearby village of Dokhin Teguri. Later the Pakrashi family showed a preference for Vaishnavism, associated with the worship of Vishnu, over Shaivism, associated with Shiva. Vaishnavism had been renewed in these parts through the efforts of Chaitanya, the sixteenth-century spiritual leader (Stewart 2012). A crude way to distinguish between the two would be to think of Shaivism as more associated with bodily practices toward martial prowess and spiritual strength, whereas Vaishnavism was more associated with the various avatars, or reincarnations, of Vishnu, and practices of devotion to Vishnu.[3] These changing inflections well depict the dominant streams of Hinduism in what was then East Bengal.

I speculate that this somewhat benign biography of the Sthal Pakrashi lineage may arise out of a degree of loyalty toward a zamindar family that actually lived and exercised their power in situ rather than in absentia as was the practice of many zamindar families, including that of Rabindranath Tagore, the famous poet who had extensive landholdings in Shahjadpur, an upazila in Sirajganj, but who preferred to be based in Calcutta in West Bengal. The familiarity produced by having the Pakrashi family close by was manifest in knowledge about the family and glorification of its fame in astrological and spiritual matters, as well as an aggrandizement of its oppressiveness communicated through many besides Nuru Ghosh. The name Pakrashi still manages to communicate a forceful presence, unlike any other zamindar in the past of Sirajganj, even though the family is long gone.

Ahmed-Siddiqi helps us to fill out the schematic account provided by Hunter on the interrelationship of Muslims and Hindus. He writes that although Sirajganj had a much larger Muslim population than Hindu, with almost 77 percent of the total population identifying as Muslim, this was still a culturally Hindu area, mainly because of the prominence of the Hindu zamindars. He says that Muslims were so identified with the Hindu zamindars that although some of them were self-sufficient landowning peasants, they still considered themselves subjects of zamindars. Ahmed-Siddiqi speculates that some of them did not even know that they were living under British rule, blaming their lack of education for their ignorance.

In an amusing and extensive list of vignettes, Ahmed-Siddiqi indicates how closely identified the Muslims were to Hindus and how lightly Islam sat on Muslim lives. He recounts that Muslims did not say their daily prayers. In fact, they said their prayers so seldom that on the festival of Eid, on the occasion of the communal prayer, they had to ask around to borrow a prayer cap. They wore the dhoti, pants associated with Hindus. They used copper plates and glasses for their food and drink, again associated with Hindus, while Muslims were apparently enjoined to use china. They insisted on oiling their bodies and taking baths before eating, a practice associated with the Hindu emphasis on purity. During Eid, Muslim villages were quiet as everyone was ignorant of the significance of the day and what to do during it, whereas they turned out in large crowds to celebrate Durga Puja and other pujas held by Hindus. And so the list continues. But by way of consolation Ahmed-Siddiqi writes that all was not bad with the Muslims in Sirajganj. After praising them for their simplicity, bigness of heart, and generous hospitality, he mentions that they supported a madrasa, a religious seminary, and that the school would eventually graduate *maulvis*, or religious leaders, who would help to rectify the erroneous ways of the Muslim in Sirajganj.[4]

The Active Absence of Hindus

The first mass exodus of Hindus happened during the partition of India in 1947. While the larger and more violent exchange of populations took place between Punjab, India, and West Pakistan, there was also an exchange of populations between West Bengal, India, and East Bengal, which became East Pakistan. As Joya Chatterji (2002, 2007) has shown, many more Hindus left for India than Muslims came to Bangladesh at that time. The next big surge coincided with the India-Pakistan War of 1965, during which time Hindus came to be indexed as sympathetic to India and as enemies within (Menski and Rahman 1988; A. Chowdhury 2009; Guhathakurta 2012). The Enemy Property Act originated from this time and was actively used to confiscate the properties of Hindus considered to be treasonous. In a classic case of a double bind, the very migration of Hindus to India to avoid state persecution triggered this law. Strangely, the third exodus of Hindus occurred after 1971, after Bangladesh had secured its independence, an issue that remains under scholarly study today.

Each of these periods of Hindu out-migration was indexed within my ethnography in land-related matters. Recall Thandu and Nannu, the two friends who fought against the Pakistani army during the Liberation War of 1971 and who returned to the village of Bishtipur after the successful creation of Bangladesh. They had a falling-out over land of unclear provenance. It would seem that each of them had a historical connection to someone who had a potton, or permission, from a different zamindar of the same land. Some of the confusion may have been caused by the early efforts by the British to make administration efficient by creating districts out of subdivisions, as they did of Pabna from Rajshahi in 1832, which by their own accounts produced multiple and overlapping sites of government and wrongful payments of revenues arising out of confusion over jurisdiction (Cotton, Burn, and Meyer 1908). Many zamindars, including Pakrashi, who was explicitly named by Thandu's nephew, played fast and loose with land, over which their claim was not entirely without ambiguity, and this may be one such case of ambiguity that continues to perpetuate confusion and conflict into the present. Thus the colonial past and zamindari machinations reached into Thandu and Nannu's present.

Sirajganj was very much part of the Liberation War of 1971 as Pakistani armies had encampments there and undertook a scorched-earth policy to drive guerrilla fighters into the open. It was also where successful guerrilla attacks were undertaken against the Pakistanis, with Sirajganj in nationalist parlance "being liberated" on November 29, 1971. However, as Nayanika Mookherjee (2015) has shown, during this period many Bengalis perpetrated violence against their neighbors as collaborators of the Pakistani army and also working independently for personal advantage under the cover of war. Thus, at the moment Thandu and Nannu found themselves together in Bishtipur, a large number of Hindus had already left Sirajganj because of the India-Pakistan War. The Enemy Property Act was active but not so active as to allow the state to know and put to use or sell every piece of property owned by Hindus. Access to settlement maps and records by laypeople meant access to an untapped potential for land, and Thandu and Nannu were alive to this possibility. Thus, their fighting an old battle over conflicting zamindari jurisdictions was interwoven with fighting over a new resource: abandoned, confiscated, and underutilized property with or without the official designation of khas or government-owned.

Over the course of this period, 1947 to 1971 and to the present, the figure of the Hindu changed from that provided by Ahmed-Siddiqi. As mentioned earlier, attempts were made to reform the Enemy Property Act through the Vested Property Act of 1974, until the latter was repealed entirely in 2001. As a consequence, Hindus were encouraged to reclaim confiscated property or at least to maintain a claim on the property they had left behind. Meghna Guhathakurta (2012) shows that few, if any, Hindus came forward to take up this promise. The repeal of the Enemy Property Act meant that chauras could not claim all land of unclear or absentee ownership as potentially khas and therefore available to be fought over and used. Rather, they had to engage in battle on the legal front, fighting ongoing cases over ancestral documents that purported to show that one's forefathers had bought land from specific Hindus or more often than not bringing forward Hindus as owners of the land with whom to engage in land transactions. Very rarely were those brought forward the actual owners. More often they were relatives with some claim to the land or with power of attorney, or distant relatives with little to no claim to the land. Usually they were simply Hindus to take the place of the original Hindus. Far from the specificity of Hindus that Ahmed-Siddiqi noted in 1916, the Hindu was now simply a generic category, absent and eminently replaceable.

I am not certain that there is a major conspiracy here. Although individual acts of obfuscation and fraud were evident, I am not convinced that Hindus were willfully dispossessed of their land and property in the chars and pushed away. If one looks at the results of the 2011 census of Bangladesh, Sirajganj showed Muslims at 80 percent of the population, an increase of only three percent from the 1920s. As Feldman and Geisler (2012) inform us, it is not the physical out-migration of people dispossessed of land that should draw our attention in Bangladesh but instead the small ways in which people are displaced in situ. In turning to my fieldwork among the chauras in the Jamuna, I am reminded that they were perpetually involved in small lateral displacements because of either floods or erosion. These displacements meant a continual throwing up into the air of the constitutive elements of social life and their continual reorganization when chauras found new ground. In this manner I watched a man go from having lush fields in front of his home, to his fields being far enough away that he needed to ride a boat to reach them, to his fields being no more, with him sitting at the threshold of his home waiting for it to go as well. I realized that it was this slow movement in place or lateral shifts that led people to just drop off the scene. In other words, the erosive tendencies of the river were as important in dispossessing Hindus of their land in this area as were wartime violence, oppressive state laws, and individual avarice.

Contemporary Hindu Perspectives

Sthal Char was above water the entire time I was carrying out fieldwork in Chauhali, Sirajganj. Its presence may have been the reason that the memory of Sthal Pakrashi was more animated than usual. Once the char broke, that memory would ebb in intensity, while other names, figures, and events would be reanimated elsewhere. If the percentage of Muslim population had increased only three percent relative to Sirajganj in the 1920s, and the Hindu population held at 18 percent, who were these Hindus, and how did they relate to the chaura and char excising of them? My first introduction to Hindus was to the Namasudras, Das, and Kapali living in and around Nagarpur town in Tangail district, which has a large Hindu population. As a reminder, the island char on which I worked and which was a part of the Chauhali upazila of Sirajganj district was wedged between Sirajganj and Tangail districts. I could see both districts from the island. The occasion for my visiting Nagarpur was the Durga Puja, and I was invited to spend the day with Shohidul's computer teacher, Gopal *da* (short for *dada*, or brother), who owned a small shop in town where he offered classes to those seeking to navigate their way around Microsoft.

I accompanied Gopal da to a *mela*, or fair, in a gigantic field in the town filled to capacity with people dressed in their best clothes, walking about and buying food, clay figures, and decorations for their hair. In the background was an enormous stage on which musical groups playing Hindu pop, folk, and classical music competed with one another, interspersed with speeches by local dignitaries. The only speeches that caught the full attention of the crowd were those by politicians from the zila, or district, and upazila levels in which Hindus were reassured that their presence and contribution to Bangladesh were appreciated by all. Otherwise, everyone was busy anticipating the moment when the enormous sculptures of Durga, both here in the mela and in individual homes, would be taken to nearby water bodies to be submerged in a ritual called the *bishorjon*.

In the evening, I was taken around from house to house in Gopal da's neighborhood to admire the beautiful life-size Hindu gods and goddesses or *thakurs* produced for the occasion of the puja. These beautiful lifelike reconstructions went beyond any thakurs I had seen in Bangladesh previously and were the products of skilled artisans coming from India to cater to the rising demand for high-quality goods among Bangladeshi Hindus. People walked around appreciating the displays in one another's homes and sat down to eat *khichuri*, a rice and lentil dish, and *mishti* (sweets). I had a moment of cognitive dissonance when I realized that what I had initially thought to be festive music blasting over the loudspeaker upon closer attention turned out to be recordings of pala gaans, or competitive songs, between famous singers from the district of Manikganj, known for its musical families. The pala gaans being broadcast debated the merits of Hindus versus Muslims. It felt odd to have this polemical discourse in which each singer mocked the premise of the other's religion in the middle of a joyous festival. Yet if one recalls that Durga Puja marks the victory of the goddess Durga over a demon and that the puja is often accompanied by mock stick battles and other competitive revelry, it makes sense to have a musical competition between Hindus and Muslims as well.

Gopal da reassured me that the pala gaan was much appreciated by its listeners, who were devoted to the music of the two famous singers, Momotaj and Rashid Sarkar: *They may speak ill about us for the duration of the song, but they also know us.* Back at home in Baltimore, I watched these pala gaans, which are readily available on YouTube, and realized the extent to which the debate between the two singers, one representing Hindus and the other Muslims, came down to two incommensurable positions. The singer representing Muslims invariably pointed to the truth of Islam as indicated by the simplicity of choices within it, the singularity of *tauhid*, or oneness, and the error of Hinduism as indicated by the bewildering complexity of its gods and rituals. The second singer invariably answered with the refrain "But where is love in all of this?" The emphasis on love was an index of Chaitanya, the sixteenth-century saint, as his emphasis on devotionalism had been crucial in renewing Hinduism in Bengal (Stewart 2012) and perhaps was a nod to the recent entry of Christian proselytizing in the area.

Generous as Gopal da was with his time, on the matter of chars and chauras he had nothing to say. Even though Nagarpur was only about an hour's bus ride from Chauhali upazila town, which was on the mainland, and many people from Sirajganj whose lands had been eroded had taken refuge in Tangail, Gopal da had no connection to this world on account of being a middle-class urban Hindu. His world was urban Tangail. The chars were some distant backwaters for him.

Locating Hindus who had once lived on lands that had subsequently become chars proved to be difficult. Everyone whom I asked in Chauhali answered that they had no idea where the Hindus who had previously lived in their villages had gone. I knew that my request was being dismissed, and it was only when I met Minnot Ali, who had recently returned from the mainland, specifically Shahjadpur, in western Sirajganj, to reclaim his land, that I was finally able to retrace the path of Hindus out of the chars to the mainland. Shahjadpur was an upazila west of Chauhali and farther in distance than Nagarpur from Chauhali, but unlike Nagarpur, it was in the same universe as Chauhali in that the river was a vivid presence among those residing there.

As soon as I got to the small town of Nogordola in Shahjadpur, Minnot Ali, our guide, became absorbed in his ring of acquaintances in a tea shop. He pointed me in the direction of a shopkeeper who hailed from Gorjan, the village south of Dokhin Teguri. Horesh introduced himself as a retired army man. He left the chars when they were still attached to the mainland, but by the time he finished his military service, his ancestral lands had eroded and re-formed as part of an island char. He shrugged to indicate that he was no longer interested in returning to land that had become char and chose to join the other Hindus from Gorjan who had moved here during his absence. He had always intended to start a business after his stint in the army, and this shop afforded the perfect opportunity. I asked if he had returned to his land in the char or if he had sold it. He said that he had never returned to see his land, and for all he knew people were cultivating it. He also knew that if he tried to return he would likely be mistreated, if not threatened, if he wasn't simply misdirected so he would get lost. However, over the years he had noticed that others who had left behind their lands had been approached regarding selling these lands. He was convinced that he too would be approached one day. This was the only way he would get some value from the land, and until then he preferred to hold on to the deeds.

Horesh walked me to the house of Chakrabarty, both a schoolteacher and a *hakim*, or doctor, who apparently still retained a strong connection to Gorjan. He was an elderly, dhoti-wearing man, immersed in devotional practices in the midafternoon, but who stopped to speak with us. Like Horesh, he had property in Gorjan. Unlike Horesh, however, he had maintained control over his property not only through his deeds but also through *borgadars*, or sharecroppers, with whom he had long-standing relations. Although he didn't have the time or wherewithal to go to Gorjan much these days, as he was old and lived alone, his borgadars apparently still visited with him to give him some share of the crops. He recounted how he was still treated with love and respect in the chars for his service as both a teacher and as what he called, in English, a "house physician" at a time when there were few doctors.

Master Chakrabarty recounted how he had to move away from Gorjan after it eroded and how he moved along an arc now familiar from my prior discussions of such arcs (see chapter 2). He moved with twenty-five families to nearby Makra when the Jamuna came westward. He bought property as he moved because he found that people were not willing to provide refuge unless he could lay definitive claim to a piece of land. In this manner he kept moving westward, trying to stay ahead of the Jamuna, buying land to secure his moves. Over time, the many families who had started off together fell away until he was no longer the person whom others followed but rather a single, elderly man with a sickly wife seeking to go where he might find some shelter among Hindus, even if they were not from his original village or jati, or caste. This was how he came to be among the Hindus in Nogordola, only a few of whom hailed from Gorjan. He would likely die here among strangers, where he had no *porichoy*, or identity, but there was no going back to the chars for him.

While Horesh the shopkeeper spoke in the voice of a victimized minority who had no standing in his original village and had been denied his paternal heritage but who still held a few chips to play in the game of property if and when he was approached, Chakrabarty Master spoke in a far more complicated, even anguished, voice. He was embedded in Gorjan's village sociality and shared relations of mutual respect and reciprocity with Muslim chauras. In fact, he went so far as to say that he was loved. He was affectionately called Nolini, with many chauras continually besieging him to return to live among them. But, as he told them, he could not do this because then who would burn him—that is, light the fire of his funeral pyre—when his time came? Apparently, chaura women assured him that they would light the fire, but he knew this would not sit well with their husbands. He acknowledged the love that moved the women of his acquaintance to make such promises, but he also referred to chauras disparagingly as "they" or "those people." At one point he also said, *When one lives in the water, one does not make enemies of crocodiles.*

Chakrabarty Master was most bitter about the fact that Muslims had taken over Hindu property in the chars, that the lands of *babus*, or zamindars, had been possessed bit by bit by Muslims secretively putting forward fake or duplicate Hindus with whom to transact to buy up unclaimed property. The Muslim desire for Hindu land had transformed a once relatively harmonious village sociality into a hornet's nest of conflict and infighting. And what had become of the Hindus who once owned the lands? The zamindars had departed long ago, with no care for what they had left behind. Others left after them. In fact, he insisted, Hindus were leaving every day. But what did they find in India? That life was as difficult there, if not more so. After all, they had no nationality or citizenship, no recognition by the state, and no family with whom to take refuge. Yet, they flourished there, he continued, contradicting himself with his next statement. Everything they took up to do, they did well, and they were rewarded with wealth and recognition.

Unable to reconcile what he was saying, I inquired after his personal landholdings. He told me that he had two plots of land of five bighas each in Gorjan, specifically Kanda Gorjan, or the Shoulder of Gorjan. The land kept coming up and going down. Right now it was up but mostly covered in reeds, the first blush of plantation on chars. Many people had already approached him when his land was underwater seeking to buy it because the prices of submerged lands are traditionally low. He refused them because, as he said, the land was going to come up no matter what, so why sell low? His plan was to hold on to his land until it was more valuable. Just as I thought that he was being a savvy businessman, he stated that he did not intend to sell because while he had the land, his name—or that of his family—was widely known. He was old and explained that he had no *waris*, or inheritor, so it was unclear what would become of his land after he passed away. Surely it would be delivered to the machinations of a property market that he despised? But he only shrugged in response to my question about this. Something about Chakrabarty Master's tone upset Shohidul, and as we concluded our conversation and walked away, he said to me, *Master sahib still did not explain if Hindus are moving away because of Muslim chauras or for some other reasons. How exactly were the Muslim chauras at fault?*

Many studies have explored how Islamization was ushered into Bengal, later Bangladesh, by Islamically inflected peasant movements (Choudhury 2001), by the rising influence of traveling maulvis, or religious figures (R. Ahmed 1982), and by contradictory state and elite practices, both colonial and postcolonial (Shehabuddin 2008). Islamization also impacted the place of Hindus in East Pakistan's Bengali society through discriminatory property laws, as I have explored earlier (Guhathakurta 2012). But few studies provide a sense of how ruptures emerged within East Bengali/Bangladeshi village communities, however agonistic they might have been (except see B. Roy 1996). How did Hindu-Muslim relations fall apart? Or, how was the memory of shared pasts subject to forgetting? In lieu of going in search of such elusive accounts from dispersed individuals, I take a Bengali film, a visit to the ruins of a zamindar compound in Nagarpur close to Chauhali, and a ghost story as providing a possible, partial archive of the unraveling of relations.

The leftist Bangladeshi filmmaker Tanvir Mokammel's *Chitra Nodir Pare* (1999) is set roughly in the period of the 1950s, when Ayub Khan's military leadership of Pakistan has rendered democracy a farce felt strongly in this small provincial town. Here resides Shoshikanto, a Hindu lawyer living in his beautiful yet crumbling ancestral home along the Chitra River with his sister and two children. His wife is present as a garlanded picture in his chamber at which he looks longingly on evenings spent listening to Rabindra Sangeet or Tagore Songs on his gramophone. At the start of the film, when Shoshikanto's children are still young, rumors are already circulating that Hindu Bengalis are migrating to India. The lawyer refuses to leave and scolds others for even perpetuating these rumors as they produce a sense of inevitability that he attempts to stanch by his dogged insistence that this is his home.

His son, Bolu, and daughter, Minu, are initially seen playing happily with the other children in the village. As his son grows older, however, and is bullied by unknown others, he refuses to stay in Pakistan any longer. He asks his father to arrange for him to go to Calcutta (Kolkata), India, to live with his uncle and study there. The scene shifts to Minu, who stays behind and is now grown up, happy in college and her home life, busy creating a relationship with a Muslim Bengali boy and bathing in the Chitra River with her aunt and friends. Eventually, however, her young beloved is killed in a police demonstration in Dacca (Dhaka), suggesting how confrontations have heated up between the students and the Pakistani military establishment. In time, others of her acquaintance are mistreated; a friend is raped by Muslim men and is later seen taking her life in the Chitra River, timing her death to coincide with the Durga Puja and the submerging of Durga in the river by worshippers. The moment captures perfectly the synchronicity of the fervor of festival, the chaos of the young girl's mind, and the choppiness of the river, with one allegorizing the other. The rumors about more and more Hindus simply leaving everything behind and walking into India consolidate into facts, and Shoshikanto finds himself in the odd situation of fighting endless land seizure and property possession cases on behalf of both their new Muslim occupants and those who have left them behind.

As the lawyer grows older and takes to ambling about somewhat confusedly as opposed to his earlier purposeful strides, it is as if his days are numbered. Indeed, one day he simply dies on the banks of the Chitra. The end of the film finds the young girl taking her aunt and leaving for India on a bus. En route to India, the aunt asks the girl in puzzlement, "How did this happen?" I take this to mean not how did we come to be out of luck but, rather, how is it that in the midst of a riverlike flow of rumors about those who were leaving we never heard the one about our own departure?

Although the Chitra River is in southwestern Bangladesh and my field site was farther north, the ancestral compound in which the lawyer lived with his attenuated family of his two children and sister was not unlike the ruins of the zamindar bari (house) that I toured in Nagarpur, Tangail, on the occasion of visiting Gopal da for Durga Puja and during follow-up visits. It was built in the nineteenth century by Jodunath Chowdhury, with Ahmed-Siddiqi's history informing us that Chowdhury was one of the prominent zamindari family names in the region. The house was majestic even in its neglected form, with a central building in a recognizable colonial architectural style, combining British imperial styles with more regional elements in its angular stone building structure, lattice-worked parapets and verandas, projected cornices, roofed balconies, decorative columns, wooden shutters, and a beautiful interplay of metalwork and inverted stone arches in the walls, festoons of metal and wood in an area where brick usually dominated as the building material. There were several buildings clustered behind the house and to its sides, but there was no one position from which to get a perspective on the entire layout, with overgrown trees, new buildings, and roads interrupting what must have once been a very grand estate.

The main building was now being occupied by a girls' college and looked to be in use as evidenced by clothes hanging out to dry and wires stretching across large open expanses to provide a thin feed of electricity to the denizens of the largely decrepit building. There were no signboards to be seen to give us any history or background of what lay around us. We later learned that the original family had moved to India, and the entire property was possessed by the government of Bangladesh, to be fitfully retrofitted for practical use or left to go back to nature. The smaller buildings, which were largely brick, appeared to be eroding, whereas the ones of stone stayed on defiantly.

I was reminded of a story Yusuf Shikdar of Boro Gorjan once told me of another Hindu zamindar, Kali Konto Biswas, who apparently deliberately refused to sell or deed his land in the chars before he left for India because he was entertained by the idea of the land breaking and emerging from the river waters only to have the nayira, a derogatory term used to refer to the circumcised penises of Muslim men, fall into infighting over it. Shikdar found this to be hilarious, loving the image of himself along with his fellow men falling over themselves over such treacherous land.

From the people milling about we gathered a few details about the Chowdhury family, such as that the son of Jodunath loved the finer things in life, starting each day with the sound of the shehnai, or clarinet, being played live, owning a small menagerie on the grounds, and sponsoring a well-known soccer club in the area. The enduring tidbits on Hindu zamindars in these parts played up their being capricious, pleasure-seeking overlords, whereas the film at least suggested that many Hindus, even of zamindari backgrounds, may likely have been ordinary professionals living in the midst of their ancestral ruins by the time of the 1947 partition or the 1965 war between Pakistan and India, or the period of the 1971 Bangladesh war of independence, the three periods that saw the greatest out-migration of Hindus from East Bengal, later East Pakistan and now Bangladesh.

When Shohidul, who was also a local boy, told me that the ghost stories that he grew up reading and hearing in these parts were invariably located in the ruins of zamindari estates, we made for a local bookstall next to the tea shop where we were grabbing breakfast and, on a whim, purchased a book titled Famous Ghost Stories, assembled by Leela Majumdar and published by a press in Kolkata, India, in 1998. Where there were plenty of local publishers and booksellers in the main city of Tangail, the presence of a Kolkata-published book in a local bookshop by the river suggested that borders between the two countries were as yet porous. We flipped through the book until we found a story that suggested such a setting of a zamindar house—"Shonkhraj" by Abdul Jabbar—and read it within a stone's throw of the Nagarpur zamindari house, treating the story as a projection of the Bengali/Bangladeshi Muslim avid imagination of Hindu zamindars. There was no way to tell when the story was written, nor was the story placed in a recognizable time period. The author's name, Abdul Jabbar, was clearly Muslim, meaning Servant of the Almighty, and he espoused similar ideas about Hindu lives as those that had already been shared with me by chauras. However, the usual stereotypes were not cast quite as fantastically as what I was hearing, but instead more ambivalently. Even the title of the story was a bit of a mystery. It was unclear whether the Shankha in the title referred to the *shankhachur* (the venomous krait snake whose action sets up the demise of the zamindari family in its estate) or to the *Yug* (era) within the Vedic astrological year titled Shankha, which only occurred at the conjunction of two particular planets and gave rise to a unique time—a time outside the usual flow of things considered a time of munificence.

The story tells of a school headmaster who has been transferred to an area and seeks comfortable lodging for his family. He hears of an abandoned zamindari bari and goes to ask its current owner, Omiyo Chowdhury, also referred to as Pagla Babu, or the Mad Sir, whether he can live in the main house, which is currently under lock and key. Pagla Babu himself lives in an adjoining building, which once housed the court over which his forefathers used to preside, along with his widowed sister, Indulekha. He warns that the house is haunted, but the headmaster is eager to live in style at a low cost. Pagla Babu also informs him that Indulekha lost her husband to a snakebite within four days of her marriage and has quite lost her mind. It was unclear what a krait was even doing in these parts, and it is hinted that the snake might have been a deity. Later the headmaster and his wife learn from others that Indulekha often walks about naked and is threatened with death by her embarrassed brother, who blames her for having brought the curse of the killed snake upon the family's name and honor. Indulekha openly claims herself to be Lord Krishna's paramour. As she says to the headmaster's wife to explain why she isn't embarrassed by her nakedness, *In my eyes there is only one male in the world. He is my Krishna. Everything else is either nature or Radha. What is shame to her! The ovary is in her* (Jabbar 1998, 152–53). In other words, with Krishna paramount to her existence, everything else is assimilated to nature and femininity. Mortal men are merely men in appearance and assimilated to Radha. Indulekha is often heard chanting "Krishna, Krishna" as she goes about her ablutions and worship.

Certain mysterious happenings in the main house, such as the sound of the shehnai being played in the middle of the night and a voice speaking to the headmaster's wife in the dark, finally convince her that the place is uninhabitable and to leave abruptly. She urges her husband to leave as well, but he makes fun of her fears. However, when he too hears the same sounds, he investigates and finds that Indulekha has for companions a monkey that can play the shehnai and a cockatoo that can mimic human voices, both of whom can engage in discourse with humans. By the time he uncovers the truth of these phenomena, he is enraptured not so much by Indulekha's beauty as by her challenge to him that he has never experienced the mystical heights of sexual union and accepts her invitation to be her Krishna. We realize that he has been seduced but also that he is finally free of conventions as indicated by the fact that he begins to take around the monkey and cockatoo to school and to openly converse with them. In the final scene, his wife arrives in a car and forcibly takes him away. Indulekha only looks up momentarily before returning to whatever she was doing, chanting "Krishna, Krishna."

One could well imagine the Nagarpur zamindar bari as the setting for this story, with the fictional family sharing the same name as the actual family, the reference to the *ghats*, or steps, leading to a pond, the animals, and the shehnai. However, these are quite generic elements that are widely shared by or attributed to many zamindars. The more important aspect of the story is that the focus is not on the profligate zamindars but on their descendants living far below the standards of their forefathers, specifically a young woman who makes her home among the living traces (the animals from the menagerie) and material debris (the physical ruins) of the past. She is an ambiguous figure, and the author espouses ambivalence toward her as to whether she is an eroticized other or the embodiment of Radha through whom one gains immersion in nature and the pleasure of union with the divine. The story also updates our picture of Vaishnavism within Bengal related by Ahmed-Siddiqi as having turned to a focus on Krishna as an avatar of Vishnu. The story could be as much about the endurance of love for Krishna amid societal disapproval as about the dangers of succumbing to this love.

This reconnoitering of a handful of cultural products—a neorealist film set in the 1950s, a historic site from the nineteenth century, and a ghost story based on one such site after it had passed its heyday—may not satisfy anyone as offering a meaningful vantage on Hindu-Muslim relations and their unraveling in the chars. I offer them regardless to show how even as Hindus leave the area and are forgotten, there still exist shards of this prior existence. This raises for me the question of when and how these shards are accessed or become animated.

Conclusion

This chapter has explored diverse ways in which Hindus and Muslims historically hung together in Sirajganj from a relatively recent past to the present, how they now view each other through a fractured lens informed by conflicts over landholdings, and what the processes of dissolution might have been as seen through the lens of a film and a ghost story.

The way in which the movements of the river appeared to collaborate in chaura projects of excising Hindus from it was quite distinct from the kind of self-opacity evident in chaura movements in relation to erosion. For one, Hindus were both forgotten, rubbed out of recollections of the chaura past and present, yet also recollected and sought out when involved in the many projects through which chauras reclaimed their lives, such as in carrying out land-related transactions. One might dismiss this as self-interested behavior, even unjust or immoral, if not evil, but everything else I have studied was no less self-interested along one plane and yet far less volitional along another. It is precisely this multidimensionality of human actions, its compulsions and ramifications along many planes of existence, that has interested me throughout the book.

Char dwellers were inclined to see their effort to secure land as bound to the fact that they had no place to go or no means to do so, enabled by the nonaction or stasis caused by the laws targeting Hindus. While they considered these misappropriations as unjust to Hindus, they could not bring themselves to see their act as evil. They were more inclined to say that nature in stasis produced chaura acts of injustice and physical decay. One could call this a naturalizing of injustice, an obviating of guilt, shame, or judgment. However, I think it is more appropriate to say that they theologized nature, considering acts that sat well with their constitution as rightful *and* natural, and those that didn't as wrongful *and* unnatural (see N. Khan 2021b).

Jodunath Chowdhury's zamindar bari in Nagarpur, Tangail. Photo by author.

Spectacular dioramas depicting the various Hindu gods and goddesses associated with Durga Puja. Photo by author.

Children dance in celebration of Durga Puja. Photo by author.

5.

Death of
Children and the
Eruption
of Myths

Eruption into Speech

ALTHOUGH A FEW SIGNS EXISTED in the chars of Sirajganj of the predominant influence of Hindus as recently as the 1970s, I generally found a widespread ignorance of that shared past among the chaura women and the young and a deliberate-feeling silence among the village matbors as to the whereabouts of the Hindus dispersed across the region. Only Wajib bhai, one of the senior matbors, bemoaned the loss of Hindus because it meant the loss of entertainment of any quality in the form of festivals, theater, fairs, and games in his village of Dokhin Teguri. But then occasionally the villagers were struck by the tragedy of the drowning death of a child, and on these occasions the bereaved mothers were moved to speak of a world next to this one in which Muslim and Hindu mythical figures associated. In a village in the southeastern part of Bangladesh, a woman told a researcher about her state of mind at the time of her son's death by drowning in a pond close to their homestead: "I thought about nothing, had no memory of anything. Now I know [boy's name] is not alive. That day I cooked with such attentiveness it was as though I was childless. After finishing, I remembered that I hadn't seen [child's name]. The other day he came to me many times, he touched the wood and searched for fish to eat. That day he did not come and I did not even notice" (Blum et al. 2009, 1723).

The article by Blum and colleagues describes how respondents subscribe to the idea that supernatural beings live in the water and seek continual sacrifices or appeasements, sending out enticements to people. Another woman said the following regarding her young daughter's drowning death: "She didn't die due to drowning. Evil took her to the water and killed her. If anyone digs a new pond they must give some rice, egg, turmeric, chira (rice concoction), nose ornaments, and coins to the pond. But we did not do it so Gongima became dissatisfied and wanted a human being in revenge" (Blum et al. 2009, 1723). It is not only women in the southeastern part of Bangladesh who speak in this way. I also heard such charges against Ganga Devi, as Gongima was also called, at my field site in Chauhali. Here she was not alone in bearing the blame. Her consort, Khwaja Khijir, also known as Khidr, was similarly believed to entice children to early watery deaths.

How, then, do we understand this unusual situation in which women explain their forgetfulness and the deaths of their children as being caused by a Hindu goddess who is no longer animated in their social milieu and by a mythic prophet within the Sufistic tradition of Islam who also is on the wane within local memory? The research study on child mortality and other such studies state that this mode of explanation is a survival of the past, one steeped in superstition. Apparently, only women provide such explanations, while men are quick to denounce them as superstitious. The study ends on the confident note that given sufficient education on the actual causes of drowning deaths, these women will be able to overcome the hold of this mode of thinking.

I suggest instead that we take more seriously the women's expressions, replete as they are with bereavement, mythological figures, and uncanny presences. In her essay "Voice as the Birth of Culture," Veena Das (1995) explores the expressions of women located in what she calls the site between life and death uttered both in the mythological register and within everyday life, such as the myth of Antigone holding forth over the dead body of her brother or the person of Asa, a vulnerable widow trying to re-create her life in the aftermath of the violence of the partition of India in 1947. Das shows how women in such spaces undertake double work, both indicting their societies for putting up their own for sacrifice and restoring the souls of the societies through their reparative work. Their work, offered up through speech, gesture, action, and silence, is the voice that ensouls society and rebirths culture, allowing for the possibility of a future.[1]

Chaura women who had lost their children in drowning deaths and who spoke as quoted here, I would claim, were in a similar space between life and death. However, their speech did not indict their society as much as suggest that something compelled them to speak. Furthermore, through their eruptive speech they brought the past, specifically that of Hindu-Muslim relations, into the present. They revealed that memory was not only hollowed out by nature that had decayed or washed away, as explored in the previous chapter, but also that which returned or was thrown back up.

In this chapter I excavate the forgotten past of Hindu-Muslim relations from the women's speech. I do so by following the mythological personas that emerge out of what they say. Drawing on Ernst Cassirer's (1953) exposition on mythical thinking, I understand a name, or more specifically the act of naming Ganga Devi and Khwaja Khijir, to be a conjuring of the real in *bhavana*, or imagination, as in the South Indian context of which David Shulman (2012) speaks.[2] This mode of approaching Hindu-Muslim relations—that is, attending to the mythological persons in their individual specificity, their reality and their encounters across religious traditions—gets us to a different stratum of their interaction than the historical or even the allegorical (Singh 2015). Here I also draw on Schelling's later writings, specifically his *Historical-critical Introduction to the Philosophy of Mythology*, in which he writes that to take mythology seriously, we have to acknowledge that mythology isn't about something else but is itself the thing that is denoted: "In consequence of the necessity with which also the *form* emerges, mythology is thoroughly actual—that is, everything in it is thus to be understood as thoroughly as mythology expresses it, not as if something else were thought, something else said" (Schelling [1842] 2012, 136). In the context of the chars, Ganga Devi and Khwaja Khijir are "not something else, do not *mean* something else, but rather *mean* only what they are" (136). In the chars, women and children emerge as singularly important in mediating between the Hindu and Islamic traditions, and between the mythological strata and the everyday. Finally, the nature that runs through these accounts is the condition of possibility for remembering and mythologizing, for rebirthing culture from its ruins, raising the question of what might yet constitute an ethics of coexistence in this place.[3]

In this chapter I follow the threads offered up by Gongima and Khwaja Khijir to the point at which the two meet and further ramify through the

lives of specific chaura children and women. I switch modes from the mythological to the naturalistic by attending to a documentary by Shaheen Dill-
Riaz titled *Sand and Water* (2004) on Jamuna char life and landscape, whose
heightened cinematic sensibility suggests how the char-river time-space
continuum offers itself to the mythic and how film serves both nature and
culture. I end with present-day transfigurations of the mythological figures
among the chauras to show how we might conceive of women's voices as both
birthing culture and putting it to rest, putting it alongside nature working
through us in life and death. My focus is on how voice births culture as nature without reducing one to the other.

The Death of Children

Bangladesh is on the largest delta of the world. It is crisscrossed by rivers, ponds, ditches, embankments, and lakes and faces monsoon rains,
floods, and cyclones. It is one of the most watery landscapes on earth,
with 7 percent of its surface underwater all of the time and two-thirds
underwater some of the time. Consequently, most people live in some
proximity to water, and, of course, those who live on chars by definition
live in the water. Drowning deaths are prevalent, particularly among the
young between the ages of one and five, and are largely attributed to the
inadequate supervision of children. Efforts are underway to have crèches
or even cribs available to young mothers—who bear most of the burden
of housework in rural homesteads, rich and poor alike—to keep children
safe (Blum et al. 2009).

   Despite the seemingly self-evident explanation for children's deaths
by drowning—that is, the ready access to standing water and inadequate
supervision—the women's responses are surprising and worth considering. Most noteworthy is the women's insistence that they had fallen into
a state of forgetfulness, even indifference, while their children were lured
to their watery deaths. This is quite unlike the case of the shantytowns of
Brazil where Nancy Scheper-Hughes (1992) has shown how high rates of
child mortality were met by a cultivated indifference in mothers toward
their deceased babies. For one, child mortality was reduced substantially
in Bangladesh over the course of the twentieth century. Furthermore,
maternal love and care for the child, by the mother caring for both herself and the baby, was encouraged from the commencement of pregnancy

(Blanchet 1984). While mothers might not be able to lavish attention on their children—and there was in fact a cultural inhibition against showing such open and preferential treatment—mothers were always mindful of their children's needs (see Trawick 1992).

The death of each child remained with the mother. Child mortality rates in the chars were undoubtedly higher than in adjoining areas on account of the lack of medical services, the difficulty of travel because of nonexistent infrastructure, and the general poverty of the population. Regardless of the frequency of deaths, those who lived on the chars deeply mourned the loss of each of their children, even the ones they were not able to know or raise, marking such deaths with milads, or prayer gatherings, that they could ill afford. So the idea that mothers fell into a state of indifference signaled less a state of their distraction wrought by busy lives and more the fact that they were compelled to forget and that they were controlled by some presence. Their insistence endured in the face of criticism by their husbands and in-laws, and it was set apart from their understanding of the deaths of children by other means. For instance, if a child did not survive infancy, the char dwellers were as likely to say that the child lacked an attachment to life as that the mother did not eat carefully when pregnant or did not feed the child nutritious food. Thus, the insistence that the children were lured by Ganga Devi or Khijir seemed to constitute a class of explanations unto itself, expressing the idea that such a death was not inevitable or accidental. A death of this type did not arise from within the child nor was it entirely due to the mother's irresponsibility; rather, it was brought about through the action of external forces. Therefore, at least in this class of explanations, there was a very strong sense of being affected not by God, the all-transcendent figure who acts on all of humanity, but by mythological figures prevalent only in the chars and strongly associated with water. The women's explanations require us to inquire about these figures and their place within the chars and the wider Bengali landscape.

Gongima in Her Place

Ganga Devi, as Gongima is also called, has a long history in Indian mythology as both a deity and, in her manifestation on earth, a natural entity—a river. As a deity she descended upon earth from heaven. As related in the Bhagawata Purana, one of several narrative texts composed between the eighth and tenth centuries CE that provide contexts for the earlier Vedic texts and the Upanishads, she was coaxed to earth by the penance of an earthly king. Lord Shiva promised that he would catch her in his hair to reduce the impact of her descent. And on earth, she became the river of all munificence, a child of the mountain and a creator in her own right, giving life and imparting immortality over the course of many civilizations into the present.

A sense of why Gongima may be associated with the drowning deaths of children in present-day Bangladesh comes from a tale in the Mahabharata, the Sanskrit epic poem written between 540 and 300 BC. In it Ganga agrees to wed the earthly King Shantanu, providing he never questions her actions.[4] As we will see later, this is how Khijir was with the prophet Moses within the Islamic tradition, as one who could not be questioned, whose actions had to be borne without challenge or one risked losing a guide to divine knowledge. As Ganga births children, she drowns them one by one in the river. Ganga's purpose in killing her children, as she explains to the silent but suffering Shantanu, is to spare the children the pain of mortality. Finally, Shantanu stops her from killing their eighth child, who grows up to acquire the title of Bhishma, or the immortal one, and to whose immortal and chaste status Khijir bears a resemblance (King 2005; Doniger 2009).

Despite Ganga's long presence in Bengal, as both a goddess and a river, one is hard-pressed to find any temples, rituals, or invocations of her in present-day Bangladesh (Darian 1978). Of course, there are few ruins of past eras extant on chars given the tendency of the landscape to erode and accrete. Yet, even when land breaks and is newly constituted, some elements of the past return or endure as witnessed by the continued hauntings by ghosts of old or places that are considered frightening to this day for having once been the site of the temple of a particularly fearsome deity, such as Kali, who has a stronger presence in Muslim Bangladesh than other Hindu deities.[5] But there is very little to mark the presence of Ganga anywhere on the mainland or on chars except in the occasional reference to her in the eruptive speech that I have been exploring.

Steven Darian presents two compelling explanations for this absence. Ganga is a deity associated with beneficence, with few dark sides. A braided river manifests her presence across the landscape. Consequently, people have felt little urge to expatiate her through construction of and worship at temples as with other gods and goddesses. Instead, they have sought to bathe, or *snana*, in her waters while still living or to have their ashes scattered over her at their deaths in order to be blessed by her.

There is another, perhaps more historical, reason. Ganga has always been associated with the trinity of Brahma, Vishnu, and Shiva. Of Shiva's role in Ganga's descent to earth we have already heard. In the Vishnu Purana, part of the Bhagawata Purana, it is Vishnu who punctures the heavenly realm of Brahma, who then releases Ganges to the earth through the hole thus created in appreciation of Vishnu's bravery. Ganga is the consort of all three male deities in diverse texts, and her fate is tied to theirs in the pantheon of gods. In iconic form she often appears atop a *makara*, or a crocodile-like figure, indicating that animals and humans equally receive her blessings. At the same time, the crocodile indicates an appetite for bloody killing and as such is associated with the death-dealing aspects of Shiva. And so it is that Ganga comes to have dark sides to her, such that immersion in her waters is not necessarily a promise of immortality but rather eternal suspension, a "copula" between life and death (Darian 1978). This is a subtle but significant distinction as it indicates that life in Ganga's waters is something other than mortal or immortal life; it is a state of being in between, a quality that is accentuated by her later association with Khijir and speaks to a particular inflection between Hinduism and Islam in this area.

Be that as it may, when Vaishnavism, with its focus on Vishnu in his numerous avatars, most specifically Krishna, eclipsed Shaivism (a tradition within Hinduism that focuses on Shiva), Ganga too was eclipsed. For Vaishnavites, Yamuna was their river of choice, with Ganga appearing only intermittently in their devotional poetry (Darian 1978), and Bengal reflected these pan-Indian trends (Nicholas 2001).

The Nath tradition that grew out of the waning of Shaivism and the rise of Vaishnavism brought these two ancient sects within itself as two pathways to *moksha*, or emancipation. David Cashin (2010) relates how in the instance of the Nathist texts dating to the medieval period in Bengal that relate the exploits of the four most illustrious *siddhas*, or yogis, they are shown to emerge from parts of Shiva's body and to be his close companions and disciples. In one instance, the goddess Gauri, a manifestation of Parvati, Shiva's favored consort, seeks the secrets of immortality from Shiva. He takes her to the shores of the lake to relate them to her in secrecy. She falls asleep while one of the *shiddas*, Mina Nath, disguises himself as a fish and listens in on Shiva's secrets.[6] Consequently, a piqued Gauri vows to test the sexual abstinence of the shiddas on account of having missed out on the secrets of immortality. She employs all manner of sexual approaches, including assuming the form of a young, beautiful woman. In this manner she is able to entrap two, Mina and Kadali, of the four, enslaving one to his desires and making the second into a sweeper, while the remaining two, Goraksa and Kanupa, are cast out as mendicants. Goraksa is particularly successful in staying Gauri's advances by thinking of her as his mother. The yogis complain to Shiva about Gauri's efforts to corrupt them, but Shiva says that while they represent the right-handed tradition of asceticism, she represents the left-handed tradition of salvation through sexual union. As we will see shortly, there are close parallels between this story and that sung of Ganga and Khijir.

Quranic Figurations of Khwaja Khijir

In his iconic form, Khijir appears as a bearded man dressed in green, standing atop a fish. The fish is likely a reference to the dead fish that the cook of the Greek King of Macedon, Alexander, saw revive when he washed them in the waters where two seas meet, alerting him to the presence of the water of life. So while Alexander seeks this water unsuccessfully, his cook, who has long been seen as a precursor to Khijir, gains immortality through drinking it. Although Khijir is not mentioned explicitly within the Quran, his presence has long been sensed in the eighteenth *sura*, or verse, called "Sura Al-Kahf" (The Cave) (Brown 1983; Netton 1992; Omar 1993). The figure of Khijir drawn from this sura fits best that of the spiritual guide within Sufi Islam. In the events recounted in the Quranic verses, the prophet Moses goes in search of the point at which two seas meet. The dried fish that he takes along with him as his meal miraculously regains life and swims away at a location where Moses finds, in God's words in translation, "one of our servants, whom we blessed with mercy, and bestowed upon him from our own knowledge." Moses asks this person, presumably Khijir, if he may travel with him so that he may learn from Khijir. To this statement, the latter replies: *You cannot stand to be with me. How can you stand that which you cannot comprehend?* Moses promises not to question Khijir's motives.

During the course of his journey alongside this mysterious person, Moses witnesses Khijir carry out inexplicable acts that fly in the face of mercy and wisdom, the very qualities bestowed on Moses by God (Omar 2010). Khijir hacks a hole in the boat of a poor man who had just provided them passage. This seemingly vindictive act is followed by an even more aberrant one of Khijir taking the life of a young boy playing among his mates. Finally, Moses witnesses Khijir restore the crumbling wall of a house whose inhabitants had just turned away the two without providing them food or refuge.

When Moses can no longer bear the injustice and cruelty of what he is witnessing, he demands an explanation from Khijir. Khijir derides him for his impatience, saying that Moses knows only how to act on the word of the law, betraying his lack of farsightedness and mystical intuition. Nonetheless, he provides Moses the explanations he seeks before dismissing him (Omar 1993). Khijir explains that the boat was best temporarily put out of operation so that it could escape the clutches of an unjust king who was impounding the boats of his subjects. The poor man who had given them passage would not have been able to survive if he had permanently lost his boat. The boy was best killed as he was going to grow up to perpetrate many evils and bring unhappiness to his parents. While they might grieve him now, they would be rewarded with better offspring. The wall was best repaired so that it might continue to safeguard the treasure hidden within it destined for two young wards of the household when they came of age. If the treasure were found before its time, their greedy guardians would surely have squandered their inheritance.

Although in each of these instances, Khijir has foreknowledge of what is to come and acts accordingly, his actions are inscrutable to Moses. And when Moses is told to accompany Khijir unquestioningly, their relationship, however short-lived, exemplifies for all time that between Sufi initiates and their spiritual guides. Within the history of Sufism, Khijir is one of the few who can initiate people to mystical knowledge by appearing to them in a dream or mystical illumination, without the intercession of a living master or the texts of masters (Omar 2010). As a figure close to God, Khijir enables not just direct access to divine knowledge but also the experience of theophany, a witness to God here robing Khijir in his divine attributes and a paradoxical retort to the problem of theodicy, the concern as to why a benign God allows evil: "In cruelty there is mercy" (Netton 1992, 14).

Khijir in Bengal/Bangladesh

Rila Mukherjee's "Putting the Rafts Out to Sea: Talking of 'Bera Bhashan'" in Bengal" (2008) speaks of a ritual, Bera Bhashan, unique and once much practiced in riparian Bengal, in which rafts made of palm or plantain leaves were pushed into the river laden with food goods. Mukherjee shows how this ritual conjoined the agrarian order with fishing and maritime activity and was centered on Khijir, whom she describes as "the patron saint of Muslims at sea" (131), introduced to the region by Arab merchants. She speculates that if *bera* implied a liminal space and *bhashan* passage, then the two together in Bera Bhashan indicated a threshold or more specifically Khijir as a threshold suturing Arabia and South Asia, as well as enabling meditation on the conjuncture of geography and cosmology (135).

If a transregional Khijir was invoked and animated through the ritual of Bera Bhashan in the sixteenth to eighteenth centuries, subsequently Bera Bhashan acquired a more regional valence. Although Khijir was the exclusive patron saint of the festival of Bera Bhashan in the sixteenth century, folklorists note the introduction of other figures, gods, and goddesses on the stage of the raft by the seventeenth and eighteenth centuries. The folklorist Muhammad Sayidur describes Bera Bhashan as having become a purely expiatory ritual by the twentieth century. He writes that it had this thrust among those of northern Bangladesh who were particularly vulnerable to the drowning of children and land erosion and who thus sought to placate the vengeful Khijir (Sayidur 1991, 3). Thus somehow the enigmatic guide of the Quran, reputed for his protective and regenerative capacities, now appeared as a figure to be feared and appeased.

In Bengal he was the equivalent of the Vedic god Varuna, the god of the waters and the lawgiver to the underwater world (Hoque 1995). Varuna is interesting because he has a dual nature.[7] As the god of the waters, he rules through caprice, requiring sacrifices to appease him. At the same time, he rules through law as the lawgiver to the watery underworld. The expiation to Khijir in the ritual of Bera Bhashan speaks to his capricious nature, while another account of him in the chars speaks to the fact that he is less a lawgiver, since within the Muslim context only God gives law and the prophets bring law to the world, and more the upholder of God's law underwater. As such, he bears Varuna's dual nature. An elderly chaura in Shaheen Dill-Riaz's 2004 documentary film best articulates this second aspect of Khijir's nature as law upholder:

> When I was a child I closely observed erosion. Before erosion begins, the water bubbles up like in a rice pot. Churning water pounds against the riverbank. The current becomes very strong. The waves create whirlpools next to the shore. Big chunks of earth crash into the water and are carried away. Our forefathers said that deep in the water lives the Prophet Khwaja Khijir. He has many laborers. They dig under the soil below until

big chunks fall down from above. We haven't seen it ourselves but it is really true. This lord of the water exists and so do his servants. In the winter months they measure how much earth they want to break away in the coming year. And when the monsoon floods come they fall upon the bank again.

In another instance of this story, I was told that Khijir only does this because God has commanded him. He measures and erodes only as much land as is required to be given in taxes to God—hence, his need for links and laborers for careful measuring. These point to the long-standing culture of land measurement for the settlement and resettlement of chars as they emerge and erode into the river waters, as well as the hated colonial practice of taxation on land. It also points to the encompassment of daily lives by the word of God, with the enigmatic and capricious Khijir as both law-giving and law-abiding.

For a less Vedic, later Brahmanic association, Khijir is also linked with the Nath traditions through the founding figure, Mina Nath, also called Matseyendranath, who assumed the form of a fish in order to overhear Shiva's secret advice on how to achieve immortality to Gauri/Parvati and by means of which Mina Nath became enlightened. Thus, through the Nath tradition, Khijir acquires another duality, that of the creativity and destructiveness of Shiva.

Ganga and Khijir: A Relationship of Involution

As we have seen, both Ganga and Khijir existed in multiplicity in the Bengal/Bangladesh milieu. With the entry of Islam to Bengal from the eighth century onward, Khijir is said to have found footing in this area through his assimilation within different series of gods and other mythological figures, be that of Bhishma, Varuna, or Mina Nath. One could say that he even belonged in a series with Ganga given their association with water, their structural resemblances as figures atop watery animals signifying immortality, their ambivalent positioning between life and death, the specific mix of the beneficial and the capricious in them, and their penchant for killing children.

Yet there is a more interesting relationship between them generative of a specific inflection between Islam and Hinduism in Bengal/Bangladesh. It is not a relationship of syncretism involving a blending of different traditions because, as Richard Eaton (1993) and Tony Stewart (2001) have shown for Bengal, Hinduism and Islam have never been combined to produce a third way. Instead, conversion has happened through the seeking of equivalence between figures across the two traditions and then either the steady eclipse of one by the other or the insertion and encasement of entirely different traditions within another. Carlo Severi (2004), writing on the Apache Ghost Dance in North America, has further shown how two traditions may exist in parallel without syncretism or even conversion through a relationship of paradox. Thus, he shows how when the shaman-messiah, who had come to lead the Apache Ghost Dance, claimed to be Jesus, he was not inserting himself into the religion of the whites, producing a new syncretic form of Jesus, or putting himself in a relationship of equivalence to Jesus. Instead, he was saying: "If I am similar to you, then I am different." And this is not simply a contradiction, but "a logical link is established between two contradicting predicates" (Severi 2004, 820). Applying this insight to the case of Khijir and Ganga prompts one to ask how it is that Khijir and Ganga, within two different traditions, not only come to be structurally equivalent, drawing the two traditions into parallelism, but also come to be identical yet different, making the relationship between Islam and Hinduism as one of involution. This relationship has ramifications for how forgotten aspects of Hinduism erupt into chaura everyday life as an element—that is, a drowning death, which is intercalated with so many aspects of both traditions—pulls that past world of interaction into the present.

During one of my early stays on the char, I accompanied a young male employee of MMS to his field site. The NGO representative was there to collect weekly payments on microcredit loans made by MMS to these group members, as well as their savings. After completing these duties, he carried out a short workshop on disaster preparedness, a central mode by which to render the chaura population more resilient to their environment but one with ambiguous effects (N. Khan 2014).

This group had recently lost a large part of their village to the ravages of the river and were expecting to lose all of it soon, leaving them no choice but to move elsewhere. I found the group to be composed of tightly knit family members who were proud of their bongsho and village. Although they were moving less than half a mile away to a place we could see with our naked eyes, they considered themselves to be moving to bidesh, or foreign lands, suggesting how tightly drawn were the lines of belonging and order, with disorder prevailing outside of them.

In the context of being asked what they do to prepare for erosion and other natural disasters, the women faithfully repeated the words of advice they had received from the NGO, which had just been refreshed by the young NGO worker. They were expected to have been keeping aside some money, food staples, and a portable mud stove to take with them should they need to move. At the same time, they recalled a festival of their childhood, the previously mentioned Bera Bhashan, during which, in a day in *Poush* (January) after the monsoon, harvest, and replanting seasons, they would float large leaves of the banana tree bedecked with flowers and sweets to expiate to Ganga Devi and Khwaja Khijir. They were insistent that it was always done for the two of them and not for one alone. They giggled, recalling the silly songs they had sung about the upcoming nuptials of the two as they pushed off their makeshift rafts:

> Dol dol doloni
> Ranga Mathaye Chironi
> Bor Ashbe Akhoni
> Niye Jabe Tokhoni

> Swing swing swing
> A comb through your bright head
> Your groom is coming at any moment
> He will take you right away

Among the songs that used to be sung during Bera Bhashan was a *panchali*, or ballad, of Ganga Devi's desire to marry Khwaja Khijir, which was sung by women and had resonances with Gauri's relations to the shiddas in the Nath tradition. In the ballad Khwaja and Madar are two orphans who grow up to become famous *pirs*, or sages. Their biography is patterned both on the introduction of Islam into Bengal through individual saintly figures (Eaton 1993) and on the competition with figures in contending traditions such as the shiddas who were rendered mendicants by Gauri's curse (Cashin 2010). Khijir acquires sovereignty of the waters. Ganga, who is the sacred manifestation of a river, burns, presumably at his intrusion into her domain. She demands that Khijir marry her because she is no longer a maiden. Khijir demurs, saying that as a fakir, or dervish, he cannot marry. Ganga takes it upon herself to trick him into marrying her and goes through many guises before deciding on a particular one of herself as a beautiful young woman in a colorful sari. As she goes in search of Khijir, she finds a child playing in the dust. When she asks after the child's parents, she learns that he is an orphan or else why would he be found on the ground and not in his mother's lap? She feels such pity for the child that she takes him upon her lap and feeds him milk from her breasts. The child transforms into Khijir. Ganga is so incensed that she hurls him to the ground, thus killing him, while she sheds tears of shame and regret. However, being a *zinda pir*, or immortal, Khijir revives, and Ganga's desires are renewed. The ballad ends with the promise that Ganga will marry Khijir on the Day of Judgment, when everyone will be assembled on the *Maidan-e Hashor*, the Great Plain, with the implication being that the entire Muslim community will witness this union. And every child will be a cherished guest at the wedding. Just as in the little ditty that the women group members sang during Bera Bhashan when they were young girls, Ganga Devi lives in anticipation of her groom's imminent arrival in the ballad. According to one particular interpretation, it is children who have died early deaths—that is, not had the chance to be wed—who will be welcomed by Khwaja Khijir and Ganga Devi at their marriage celebrations (Sayidur 1991).[8]

In the Nath tradition, which has a similar story of the pairing of Gauri with Goraksa Nath, we are given to understand that the two represent different pathways to salvation. However, in the case of these songs, the relations are presented as more agonistic, with each seeking to assert control over the other. But even these more aggressive moves enact different relations of encompassment, with Ganga going from seeking dominion over Khijir, who has trespassed on her territory, to being joined with him through the sharing of milk, to attempting to kill him, to seeking to satisfy her desire for him through matrimony. These involutions produce Khijir as a direct agonist and object of desire of Ganga, but also one begotten by Ganga through being nursed at her breasts. Can one think of this relationship as exemplifying the difference in identity that Severi outlined—namely, "If I am similar to you, I am different"?

I would further speculate that it is this relationship to Ganga that allows Khijir not only to have mythological figurations in Bengal/Bangladesh but also to be naturalized as a part of the physical landscape, as being literally within Ganga in her manifestation as a river. As Steven Darian (1978) writes about Ganga in Bengal and Bangladesh, within the Marufati, Murshidi, and Baul mystical poetry and songs, Ganga is recollected not as a divine figure but, simply, as a river of contradictory nature, an archetype for all other rivers. Might then Khijir's identity, yet separateness from her, introduce new strains and tendencies within the river? Ganga threw her children begotten with Shantanu into the water to save them the pain of mortality, but, as Darian gives us to understand, immersion in Ganga's water provides not the assurance of immortality but rather a suspension from life and death. Khijir in the presence of Moses killed a child because of the cruelty that the child would wreak on others once he grew up. What did the bringing together of the two portend for children who had died in drowning deaths?

If we were to ask ourselves again what the bringing together of Gongima and Khijir meant for children who had died from drowning deaths, we might now readily answer that it meant their deaths were neither accidental nor willed by God. To remind ourselves again of Ganga and Khijir at the scenes of the death of children, Ganga drowned her own children to save them from the pain of mortality. In other words, she sought to save them from the cycle of samsara, or the cycle of death and rebirth. Khijir killed a boy playing with his mates because he too wanted to save the child from the pain of mortality, which in the boy's instance meant fulfilling his *qudrat*, or destiny, to be a cruel person causing suffering to many and bound to a terrible punishment upon his death. Neither Ganga nor Khijir offered immortality; they offered death as a release from samsara and qudrat as freedom. Ganga, with her dark sides, offered release from the categories of both mortality and immortality. And Khijir provided a particular answer to the question of theodicy with "In cruelty there is mercy."

Muhammad Iqbal, the "Poet of the East," who was born in the Punjab, the land of five rivers with which Bangladesh was once affiliated in undivided India and later in Pakistan, writes of interrogating Khijir in the poem titled "Khidr the Guide" in *Bang-e Dara* (Call of the road), published in 1924. His retelling of the story of Khijir would sit well alongside chaura women's understandings of watery deaths as being more than they seemed:

> To your world-ranging eye is visible the storm
> Whose fury yet lies in tranquil sleep under the sea
> That innocent life—that poor man's boat—that wall of the orphan
> Taught Moses' wisdom to stand before yours wonderingly.
> (M. Iqbal [1924] 2003, 40)

Khijir's prescience is to be able to see what furies lie beneath calm surfaces and to alight upon them toward intensifying them. Iqbal reads Khijir's actions as interventions into the eddies of time, as wresting the truth of life from time:

You shun abodes, for desert-roaming, for ways that know.
No day or night, from yesterdays and to-morrows free
—What is the riddle of life? (41)

As in the Quran, this Khijir also does not disappoint. He readily gives a reply to this riddle:

Constant circulation makes the cup of life more durable
O Ignorant One! This is the very secret of life's immortality. (42)

Thus, only in movement is life possible. And the perception and appreciation of movement in the midst of apparent stillness, including death, is Khijir's special attribute (see also Omar 2004).

The Lives of Children

Shopon was eight years of age when I first met him and was almost fifteen when I ceased returning to the char regularly in 2017. He was my friend and lived with his family on a *bhita*, or upraised household, kitty-corner from my own room in the MMS offices at Dokhin Teguri. That he was very trouble-some to his parents was apparent early on as I woke most mornings to his mother calling to him, shrieking that he needed to get his work done, be it cutting grass for their animals, eating his breakfast, or bringing water. He was among the throng of kids fascinated with my presence early on during my stay and among the few who lingered on after my stay extended into months and years. He refused food when I offered it to him to underline the fact that he was drawn to me for my own sake and not for any crumbs from my table, but he was always happy to receive candy, which was a genuine treat. He realized that I was there to do *pechal*—talk rubbish—and he was happy to join along, interjecting stories of his own while I chatted with others, following me around to tell me stories and describe events of which he felt I should take note, and even on occasion surreptitiously following me off the island without my knowledge just to keep an eye on me and to have some adventures of his own.

Shopon served as a key informant, and it was through him I realized that the village of Dokhin Teguri, even in its relatively new form, had haunted areas that had carried over from previous versions of itself. He also told me of the many supernatural beings, *bhut* (ghosts), *petni* (witches), and *pari* (fairies), that beset the young, speaking to them in thin, shrill voices, beckoning them to play by the waterside in order to kill them. But, as he always reminded his friends, the beings could always be told apart from humans because their feet never rested on the earth, instead gliding above it. He had saved himself from sure death several times by taking note of the feet of the young children who called out to him to come play. He knew of Gongima, although less so of Khwaja Khijir, and knew that her purview extended beyond that of the other supernatural beings who presided over small puddles, ponds of water, and the riverside. It was through him that I realized that not all premature deaths could be understood as a form of release as in the case of watery deaths.

Halfway through my fieldwork, in 2013, I started to hear about the rise in incidence of lightning deaths in the chars. The monsoon season had been disappointing for several years, with the usual cloudy skies not accompanied by regular showers. Instead, the sky stayed dark and rumbled ominously. Lightning was striking those caught unawares in the middle of the open landscapes of the char. Previously, such deaths were so rare as to elicit the charge of kalo jadu, or black magic, but now their increase had led Shohidul and his mother to speculate whether it wasn't caused by the warming of the earth, the increased production of clouds, and the friction between them as was regularly being reported over Bangladeshi radio. Climate change was in the air. However, Shopon sidled up to me one evening to speculate that climate change may be the cause of lightning deaths, with the deaths occurring randomly, affecting people who were by all appearances good, but the deaths surely served someone's cause? He recounted that whereas once the bones of those who had been struck by lightning were much sought by gravedigging magicians who wanted to harness the power of the lightning, the magnetized bones were now being sought by scientists. These bones served as excellent radars (pronounced rah-darrs), and the scientists wanted them to detect and stop planes.

I was skeptical, thinking that Shopon was pulling my leg, but then I heard the same story repeated by adults. By the following summer it seemed to have become established as the truth, with people pouring cement over the graves of their family members who had died of lightning strikes to ensure that scientists would not rob their bones. It is beyond the scope of this chapter to investigate these deaths in their own right, but I mention them as a category of death that did not lend itself to a mythological or theological reading. Lightning deaths were simply unnatural and an instance of the growing predatory nature of the wider world. I also provide this account as it was told to me by a child who was reflecting on the many forces preying on his life and world.

The Voice of Women

How do we understand the eruption in the women's speech with which I began this chapter? What might constitute the voice of women within the context of the drowning deaths of their children? To explore this question we first need to have an appreciation of the general situation of women in the chars. Chaura women were subjected to considerable male control and surveillance. It was not simply that women were the means to assert order in lives given to continual disorder, but rather the control of them arose out of the desperate sense of extending some modicum of protection to one's family. This assertion of control over women's movement and sexuality existed alongside the usual abandonment of women by their husbands, a strange juxtaposition of anxious monitoring and wanton uncaring (Westergaard and Hossain 2005).

Women were embedded in this order and central to its reproduction. For instance, some women had never set foot in the tiny marketplaces that abutted their homes despite their need for the occasional food item or daily provisions because the norm was that women did not go to marketplaces even during the daytime. And, as described in previous chapters, women were very much involved in all aspects of chaura lives, including fighting over land, making the necessary arrangements in the instance of erosions, electioneering to keep their villages intact, and even collaborating in the active forgetting of Hindus with whom they once shared lives.

In her ethnography of Bangladeshi Muslim women's subjectivity in a village in Faridpur district, farther south than my field site, Jitka Kotalova (1993) presents them as "belonging to others," as conveyed by the title of her book. This subjectivity is rendered beautifully in her writing, which reveals what it means to be the product of one's relations to others, necessarily subordinate and submissive without necessarily entailing the loss of self. A selfhood in which one's speech is always overwritten by others is not simply one of oppression; it is one in which one's speech is also always interwoven with doubling, self-doubts, and the occasional questioning of injustice against the self. As mentioned in chapter 2, it took me a while to realize that although chaura women did not have linear narratives of their experiences of displacement caused by erosion, by following their ellipses of memory, side stories, and vague recounting, I was able to understand how they viewed themselves in two places at the same time—that is, their parental home and their conjugal home—such that they endured double erosions by suffering with their natal family alongside their own. Consequently, their recounting reflected this experience and produced no end of confusion for me until I learned to listen to these narratives. Furthermore, as James Wilce (1998) has shown in his magisterial study of complaint talk among rural women in the Chandpur district of southern Bangladesh, they speak a lot all the time but in a context in which their speech is often unheard or even undermined. This deafness toward their words produces self-doubt and uncertainty about what one knows or claims, but on other occasions it also enhances the sense of wrongdoing against oneself.

While char women's subjectivity was that of being the object of others and their speech was similarly written over by other voices, their actions and gestures sometimes exceeded their speech, producing a situation in which they seemed to be doing more than they were saying. Soneka was an elderly woman living in Dokhin Teguri with whom I got along very well in part because of her crackling sense of humor and her avid rendering of the changes that time had brought to the village milieu.[9]

Soneka's situation illustrated some of the ways in which chars and their shifting movements led to confusion regarding one's status vis-à-vis one's natal home. When Soneka married someone from Dokhin Teguri, she moved here and remained until her husband died. At this point she effectively became an unprotected, helpless widow with several young children as her dependents. She sought to return to the home of her father, now her brother's home, but when that natal home eroded, she brought her brother, Shukkur Member of Chaluhara, to Dokhin Teguri to take refuge there. At this point her status went from being the wife of someone to being the widowed sister of Shukkur Member. So her affinal and natal homes were now crosshatched, brought together by the movements of land.

Soneka remained poor, but with the modicum of protection that her brother could afford her from salacious men, moneylenders, and village leaders, she managed to raise her children to the point that they could look after her. She now lived with her sons and her daughters-in-law on land of unclear ownership, possibly khas, or government-owned. She had only one daughter whom she had given in marriage to a man in the village who was reputed to be beating her regularly. When I asked Soneka what she thought of this, she said that she had been lucky to have had a husband who was a prince among men, who respected her too much to raise his hand against her, but if one was not destined for such a man, it was women's lot to be beaten.

Soneka recollected a *hastor*, or saying, that had fallen into disuse but that suggested women were doing more through their actions than what they said or understood themselves to be doing. She remembered a time when water was not so plentiful as it is today, now being just a pump of the tube well handle away. During her *boyu kal* (her life as a young bride) she had to bring water from faraway wells and rivers to help her mother-in-law with household chores. When only a cup or so of water remained, she would be tempted to throw some lentils into it to wash them so as to put to use the last bit of the precious water. However, her mother-in-law would stop her, warning that throwing in the lentils presaged the drowning of her children in water and forcing her to throw away the water. In an interesting way, the gesture of Soneka throwing lentils into water made her structurally equivalent to Ganga throwing her seven children into the water, demonstrating the sedimentation of a mythical strata within everyday life and the inadvertent resonance with that strata produced by one's acts.[10]

Another way to think of the interdiction upon Soneka was that the lentils in the water were a figuration of drowning children, a miniature of helpless bodies in water, and of the distracted purposiveness of the mother in putting them there in the first place while she busied herself with housework. It was an indictment of the mothers for the drowning deaths of their children. In another instance, I heard the women of a household blame a neighbor for the drowning deaths of her three sons because she had not taught them to swim when they were babies, followed up by saying that she had gone so far as to feed them milk from a feeder rather than from her breasts. These were both taken as signs not only of her maternal overindulgence but also of her forgetting her place in the world—at the bottom of the social hierarchy for being a chaura woman for sure, but, simultaneously, in the middle of the river. Furthermore, in their article Blum et al. (2009) explain that customarily a woman was not allowed to attempt to save her drowning child for fear of being bewitched. The claim was that the body was already enchanted and would contaminate her. Given that it was considered acceptable for others to attempt to save the child, the fear might have been that the mother might make further attempts to drown the child, deepening the association of mothering with inscrutable but likely murderous instincts.

It is within such a context of patriarchal encompassment of the social that the bereaved women spoke of the strange forgetfulness that befell them during the event of their children's drowning. There were two aspects of this speech that I want to draw out to suggest how it constituted voice in Das's sense of indicting society, providing a reminder of wrongs done but also restoring the soul of society through reparative work. Despite the weight of judgment against them, the women insisted that the deaths were not their fault but that of Gongima and Khwaja Khijir. In so insisting, they were not intransigently denying culpability or resisting societal evaluations but rather speaking with some authority, indicating a region in their collective experience in which there was agreement that certain deaths were indeed caused by malevolent forces in the water and that those forces had particular mythological figurations. Sandra Laugier (2015) identifies such agreement as a criterion for speech that claims to make a general, but not a universal, claim. For the women to speak so, they were insisting on, even perhaps resurrecting, this agreement, and in effect drawing into existence a past in which mythical beings associated with each other across traditions. In this manner, they reminded their society of its link to the past and re-souled the society.

But yet another aspect of their recollections of their state of mind during the time of drowning is important to consider. Despite the judgment levied at them, they insisted that they were in effect enchanted during the event. Their profound distractedness was not produced by the amount of work they had to get done; it was a profound forgetting that entailed forgetting that they even had children. In chapter 2, I explored the aspect of nature as the unconscious that runs through chaura actions in the face of erosion. I speculated that this obscurity to oneself was a trace of the workings of the river on the chauras, rendering them into sediment, entraining them to move in a certain direction and collect at specific spots. Of course the river was not the only playing field within which the chauras were located. In addition to the force exerted by the river on their movements, they were also buffeted by political forces, economic opportunities, patron-client relations, and kinship ties, but they were very cognizant of these forces. I speculated that in the instance of erosion there was possibly a suspension of the self to enable the river to work through an individual. Might one similarly speculate that the absorption in their work and the state of forgetting insisted on by the women during the time of the drowning deaths of their children were marks not necessarily of the river on them, as they were at some distance from the water during the drowning event, but of nature more broadly conceived? Thus far in the preceding chapters we have contended with the tendency of matter to seek materialization in legal forms and physical fights, the river's erosion expressed by the unconscious movement of chaura bodies, the construction in one's mind's eye of the ecosystem in which one is embedded and participation in this larger system, and in the ruination of the memory of a shared past with Hindus in collaboration with the physical decay of nature. Might one then understand this unconsciousness at the heart of the women's experiences, followed by their compulsion to narrate, as speaking to the tendency of nature to protect by drawing the women into an abyss of unawareness, and to rebirth by compelling them to recall a mythical past? The women's receptivity to this arc and their insistence on it gives voice not just to a past of shared lives with Hindus but also to nature resuscitating and reproducing culture from its ruins.

The Call to Mythical Thinking

Although it took the occasion of the drowning deaths of children for Gongima and Khwaja Khijir to be recollected, and with them a forgotten past to be resurrected, in actuality the past saturated the present. In particular it made everyday life in the chars highly resonant with the mythological, particularly in its world-destroying and world-creating figurations and tendencies. In a conversation with the filmmaker Dil-Riaz, I learned that he felt compelled to bring a cinematographic sensibility to his documentary on the Jamuna chars because otherwise he was unable to present the thick press of reality on the ground, which well-nigh brimmed with more presences than he could document.[11] In this section I describe several scenes from the documentary to explore the press of presences around children and women that make the scenes hum with mythic potential. They also suggest the extent to which nature lends itself to mythologizing, which is another way of thinking of my proposition that nature creates culture, that there is a relationship of identity between them.

The documentary *Sand and Water* is a meditation on the rhythms introduced by the Jamuna River upon the lives of those who live in it. The title refers to the sharp contrast between the winter and summer seasons, with winter marked by the sand beds left by a retreating river and the summers marked by an overabundance of water, perhaps even floodwaters. The documentary begins with the filmmaker on a boat winding his way between houses in a flooded village. We are offered shots into houses filled with water in which women sit atop their beds, the only upraised and dry surface. Their gesture is that of waiting in weary repose. The other extreme that we are offered is of sandy dunes, which seem like desert islands but which we quickly realize are teeming with mercantile activity. In the few instances we see of women traversing these sandy landscapes, they complain volubly about having to walk on sand, which causes them to slip sideways rather than take steps that propel them forward. Both landscapes are relatively quiet, as scale muffles the usual sounds of human settlement and interaction, but we hear sobs float across the landscape as women sit bemoaning their fate of having lost their houses and land to floodwaters.

In the scenes of flooded villages, we see many children. One is fully submerged in water, with only his eyes and nose upraised as he looks curiously at the filmmaker on the boat. In other scenes children jump delightedly into the water as their guardians ruefully shake their heads and tell the filmmaker it is impossible to keep the children from the water. And in one scene in particular, one is reminded that joy and tragedy, threat and pleasure mutually inflect each other—that Gongima and Khijir are close by. In this scene, the filmmaker seems to have kept his camera running, capturing children playing in a pond in the rain. They have long forgotten about his camera trained on them. Among the children playing, one girl, perhaps eight years of age, is evidently the caretaker of two others, perhaps five and two. She takes them ashore, putting them securely on the land before diving back to play with her friends. The two-year-old is bereft and immediately dives into the water. The five-year-old jumps in, struggling to bring the youngest back. The eight-year-old must have caught sight of what was happening as one can hear her yelping and thrashing off-screen to reach her two wards. She struggles with the two younger children and drags them out of the water, with the two-year-old holding on to her with obvious delight. In the final credits we learn that of all the children we have seen over the course of the film, the one whose eyes had earlier tracked the film crew had died (through no fault of the filmmaking process). Also among those who died was the elderly chaura who related the story of Khwaja Khijir that I recounted earlier, in which he explains that the lord of the water exists and that he is bound by God to undertake his grim tasks.

In one other scene, a woman in a household at which the filmmaker rests is working on the task of killing, scaling, gutting, and cooking fish. While the camera peers over her shoulder, she carefully tends to her work but also attends to the animals that throng her. She throws the cat some bones and she throws the dog the guts, mindful of which animal can eat what. The camera lingers on her hands as she cuts, adds spices, and then throws the prepared fish into a metal pot over a mud stove with a smoky flame under it. She has to attend to the flames the entire time as there is a breeze wending its way around the courtyard that keeps breaking the flames. The fish cooked, she quickly distributes it onto tin plates before turning her back to the eaters while she swishes ashes into the pot in preparation for washing it later. It is not clear whether she has saved some food for herself, but as I watched that scene, I felt as if she might be a modern-day version of King Alexander's cook, Khijir's precursor, who had figured out how to make a fish deliver life and immortality. The camera in effect unfurled the cosmogony coiled within the woman's actions.

Conclusion

This chapter began with a description of puzzling encounters with women who had lost their children to drowning deaths. They blamed the deaths on the evil of Gongima but made no mention of Khijir. The women in the char spoke of Khijir in the past tense, as one whom they used to invoke during Bera Bhashan but not since this festival had lapsed. In only a few places in Bangladesh was the festival still undertaken by specific families, but this too will likely pass.

When I directly asked more women about Khijir, they professed not to know anything about him, not in the way they knew about Manik pir, a figure associated with the care of cows (see S. J. Ahmed 2009, 2010). It is worth noting that cows had made a new entry into the lives of the chauras, specifically women, who tend to them in their homesteads. If the women recalled a watery figure at all, they spoke of Mayicha Dayo, a scaly being who lusted after fish (his name literally translates as "Give Me Fish"). He was said to roll onto fishing boats to lie next to fishermen whispering entreaties in his ghastly, unearthly, watery voice, or he crept as far as he could onto land to extend the stub of a limb to women to ask them for a share of the fish they were cooking, suggesting a different kind of haunting by lost children.

I was puzzled, since Khijir ramified in so many directions across time and space from the many expressions and gestures, however fleeting, I had harvested from the char. I asked Salam, the fisherman's mother, who was one of the oldest chauras I knew (although I didn't entirely believe that she was a hundred years old) and who usually had many fantastical stories to tell about possessions by *jinn*—God's creations of fire—of men taken to the land of fairies and of Mayicha Dayo, what had become of Khijir? *He comes from the Book* [a reference to the Quran] *and that is where he lives. We common folk don't have the book knowledge or courage to call on him.* So it would seem that all the while I was sensing his presence in the char, he was ebbing away, at least among the women.

A self-described lover of Khijir in the char, a well-kept man for these parts, who was dressed in white garb with a bright green turban atop his head of oiled coils of hair, provided me with two sermons on Khijir. These suggested how a Quranic version of Khijir was being newly taken up in the chars, providing another vantage on the women's association of him with Gongima. The lover of Khijir had heard these sermons from his spiritual guide, a locally famed pir of Sabri Chistiya affiliation named Maulana Nechari. These were his recollections of the maulana's sermons, which underlined Khijir in his Quranic aspect as guide to the Sufi knowledge of the invisible:

> A man goes on a journey with Khijir as he wants Khijir to open his inner eye. Khijir bids him to go to the nearest market to seek out the only man there. He wonders what Khijir means as men belong to markets and wouldn't they be there in large numbers? But when he gets to the market he finds that it is crowded with dogs, dogs barking at other dogs, dogs conducting the business of buying and selling, and so on. He races around the market until, finally, he sees a man. He approaches the man and inquires after him. The man is grumpy in his answers but says that he is unafraid to give bold answers as he has nothing to fear, for he has divided up his wealth carefully to give one-third to charity, one-third to his family, keeping one-third for himself (this being the usual formula for zakat, or charity). So in meeting this man, the first man realizes that Khijir had indeed given him the inner eye to make the real man visible to him.

In a second, more convoluted sermon, a child wants to pursue higher education to become a religious scholar. His family, particularly his father, is cruel to him, forcing him to tend their farm animals. He weeps to his mother, who directs him to go sit on the road. There he encounters a person like no other, green in dress with a green turban and flashing green eyes, who asks him what he seeks. When he tells of his aspiration to pursue religious training, the man shows him the road to a Sufi lodge in India where he can get an education. His mother identifies the man in green as Khijir and sends her son out into the night to pursue his studies. However, life in the lodge is not easy. He is again bound to a situation of tending cows and goats. When he wearies of ever learning anything, the head teacher bids him to teach his master class. He is too ashamed as he knows nothing, but the teacher draws him to his chest and in so doing transforms him into a scholar. He is henceforth able to teach and so returns home.

These sermons, which bear structural resemblances to the Quranic figuration of Khijir as a Sufi guide, may indicate the onset of Islamization in northern Bangladesh. Yet there has always been an attraction to the reformist zeal within Islam in these parts (Choudhury 2001). In addition, the sermons may be making visible or rationalizing the long-standing importance of esoteric, or *marufati*, knowledge that was also intrinsic to these parts (Hatley 2007). This too is nothing new. Curiously, however, although the sermons showcased guidance that came unexpectedly, self-revelation, and the working of miracles, the feminine was in short supply within them. This might be because Khijir was so resolutely male in these expressions, not capricious like Varuna, or a devotee of God undertaking his lawful destruction for him, or a child on the lap of Ganga, or Mina Nath secreting himself between a deity and his consort. In these sermons he was only ever on dry land, secure in its identity as the marketplace or the road to the Sufi lodge in India. If previously it was Khijir who introduced the possibility of a paradoxical response to the drowning deaths of children, in considering that in cruelty lies mercy, it now appeared that it might have been Ganga or the women all along, those living on uncertain grounds, who intensified these qualities within Khijir. He needed them. So continued the loop of interpretation between nature, humans, and their mutual crafting of culture, with nature as the ground of culture.

The bereft women of whom I speak appeared to me to be articulating this realization in the mythological register. Even as they deeply mourned their children, they couldn't shake off the feeling that death was essential to their milieu. As I write these words, I can hear a line from a song I heard in my childhood, from performers on television solemnly singing the 1940s song written by Mohini Chowdhury:[12]

Prithibi amarey chay
Rekhona bendhe amay
Khuley dao Priya
Khuley dao bahdor.

[The earth/world] desires me,
Don't keep me tied
Open [the knot] my dearest
Open [the knot] my brave one.

The figure of Prithvi (Prithibi) in this line from the song is interesting because it holds together the image of the earth in its materiality, as soil, land, or ground, with the world in a more ideational, desiring sense. The appearance of Prithvi suggests that a turn toward death is a turn not only toward oblivion but also toward the embrace of the earth.[13]

China Khatoon cuts fish with her *dao* (cutting instrument). Photo by author.

(Below) Surjo Banu fishes in the pond next to her house. Photo by author.

(Opposite) Khokon insisted on being interviewed in the same way as others, with both participants sitting on chairs. Photo by author.

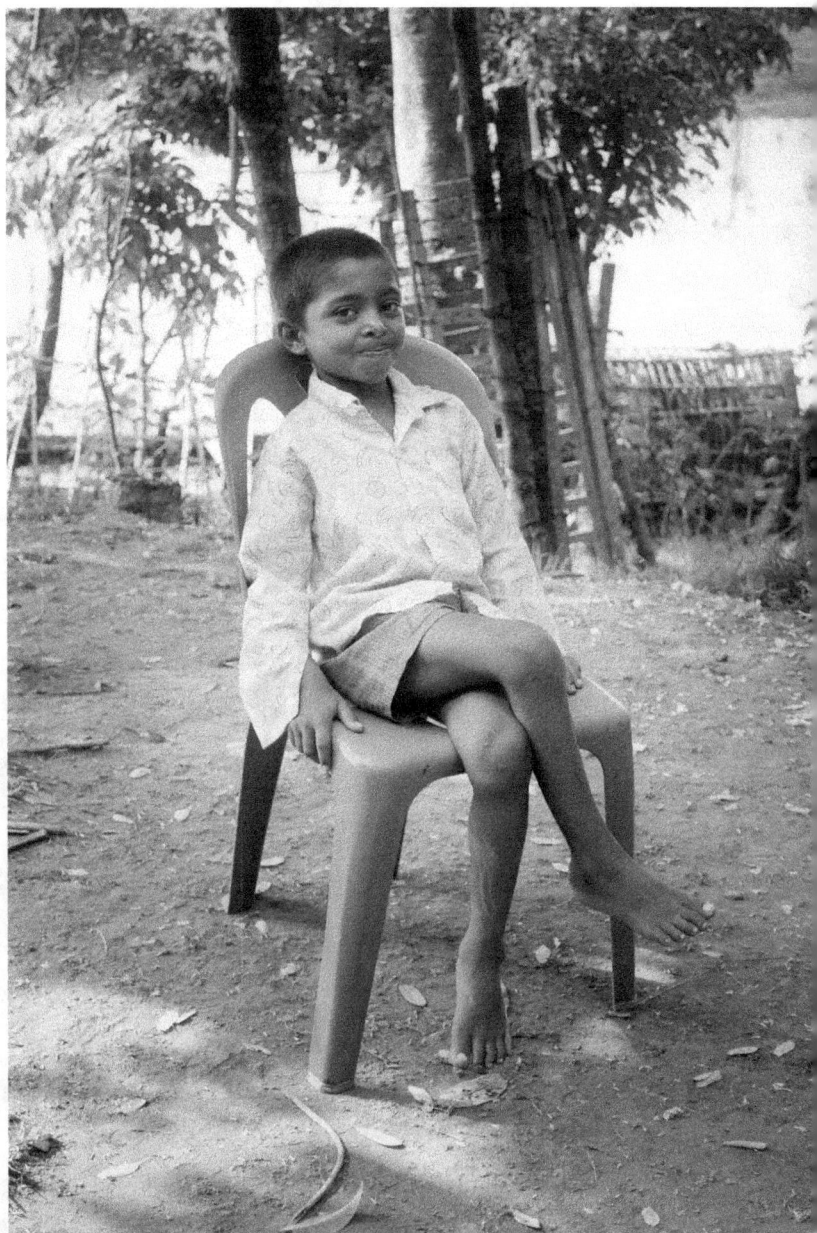

A lover of Khwaja Khijir walking through the market wearing a green turban and a white tunic. Photo by author.

EPILOGUE.   The Chars

                                in Recent Years

I WENT BACK IN SUMMER OF 2019 after I had submitted my manuscript for
review to spend a few days with friends in the char. Much had changed—
but, then again, not that much. The MMS main office off the primary road
connecting Dhaka to the northern part of Bangladesh had been the largest
building in the area when I first visited it in 2010. Now it was overshadowed
by an enormous structure, a new power plant that the Chinese had built
and that now hosted many Chinese technicians and operators who came to
play badminton with us at the MMS compound. The power plant, a multi-
lateral project among Bangladesh, China, and the World Bank, was an intimidating building with high security, but everyone at MMS was generally
cheerful about it because it meant that foreign investment was coming not
just to Bangladesh but also outside of Dhaka to its outlying areas. The path
to the char in Chauhali was pretty much the same, requiring a car, a boat,
and several motorcycles.

        I felt a wave of familiarity and a gush of emotion as I returned to the
MMS office in Dokhin Teguri, my home away from home for so long. It too
was somewhat built up. A long meetinghouse of sorts had come up on the
one open side of the courtyard that had previously afforded a view of the
fields and houses beyond. The most shocking thing about this structure
was that it was electrified, affording us lights and also ceiling fans to keep

the oppressive heat in the tin structure from bearing down on us. I learned that private companies had started to offer solar-produced electricity on a mass basis on the chars, affording many with something that had previously come only intermittently and unreliably from batteries, generators, and individual setups for solar energy.

Just the addition of electricity to this milieu already made it something other than what I had experienced. However, I was given little time to mull on the difference before I was rushed off to be put on display in a small informal school and tutoring center that MMS had set up in my name after I left in 2012, at the end of my yearlong fieldwork. I remembered a time at the beginning of my fieldwork when I startled a young girl into noisy tears when I simply addressed her, the experience of an alien speaking to her being too much for her sense of personal safety. But this time the children addressed me, peppering me with questions about where I had come from, what I was up to, and so forth. And I was shown a computer room that had been added to the center in which two older girls in starched uniforms smiled shyly at me while proudly showing me how they were learning to operate the machines. Again, it was hard for me to wrap my head around these changes, because of course I had loved Dokhin Teguri as it had been. Being uncomfortable with the aggressive developmentalist ideology that seemed to inform the mainland, I had appreciated the particular modulations of that ideology within the chars. But how could I not appreciate the effort to educate girls in my name or to use me as an exemplar or prod?

I had taken my father along on this visit, so all the men who had been gathered to speak with me, such as Moinul and many others whom I did not know, collected around him to speak to a *murubbi* (guardian). I was given free rein to go speak with Kohinoor, who had been tasked to cook for me one more time; with Jahanara, the regular cook consigned as her assistant; and with Momotaj, Morium, Soneka, Nurjahan, and others, who had been invited to eat with us. Momotaj, mother of the tea stall owner Muktar, told me that she had moved away from Dokhin Teguri for good to set up a home with her middle son, Peskar, and his wife, Shimul, in the middle of Sthal Char. She felt she was more needed by them than by her other sons. I was shocked to find Morium with the other women, given that Rihayi Kawliya/Phulhara was a ways off and her heart condition didn't allow her to travel far, but then I found out that she was now living close to Dokhin Teguri because Rihayi Kawliya/Phulhara had completely eroded in the intervening years. All my friends who used to live in far-flung reaches of the char were pretty much clustered around Dokhin Teguri. Soneka was her usual smiling self, showing off the loss of yet one more tooth. I burst into tears speaking with Nurjahan, remembering her husband, Shariat Kha, whose boat I often rode in my trips to and from the mainland and who had been such a lovely interlocutor to me. I had heard of his death while in the States. When Rihayi Kawliya/Phulhara broke, he and his family had moved to the mainland, where he had taken up driving a small van. He died in a car accident, reminding me of an early impression that I had that chauras were more often vulnerable to the caprices of mainland activities than to nature's excesses within the chars. We sat around exchanging stories of people we knew, while occasionally breaking off to embrace. At various points some of my favorite male interlocutors, such as Salam, who ferried Shohidul and me around, came by to squeeze my hand, and Shopon, now grown to my height, came to give me his salutations. I had to go in search of the ever-elusive Shontesh, who was sitting with his wife engaged in another one of his schemes to make some money—this time, making samosas to sell in the local market.

It was a short visit but enough to give me a sense of what perdured and what had changed among my friends. But I still felt that my efforts to show that the chauras, the land, and the river waters were expressions of a nature understood less as matter and more as tendencies, forces, or rhythms in the world held and remained relevant. My effort was to invite us to attend to nature in more ideational, less material ways, if only to take a break from our dogged commitment to the overly substantialized body and environment in our current theories of embodiment, vital materialism, and geo-ontology, rehearsed in the introduction.

At the same time, I also attended to the colonial and postcolonial history of Bangladesh that made the chars a necessary place of habitation and the chauras vulnerable in particular ways, subject to a particular development logic and aid projects. I attempted to draw attention to the political economy of land and its imbrication with Bengali Muslim kinship. I unfurled a schema of nationalist narratives that privileges floods over erosion and other natural events as the marker of time and legitimizer of suffering. I showed how the village as an administrative and political unit fares as an index of self-government for Bangladesh's international donors, not unlike colonial times. I explored the mechanisms by which Hindus were displaced and dispossessed of their lands within the chars and how they were simultaneously disremembered. And I excavated the ways in which this rubbed-out past perdured into the present even if only through occasional bursts of memory and in mythological forms. In other words, I sought to show nature acting through historically specific concrete lives.

I turned to Friedrich Wilhelm Joseph von Schelling and his fellow romantics in their project to find nature in themselves after its perceived separation from them and drew on their particular angles of entry into nature, through the mind and its attraction to organization in matter, through the unconscious, through leaps of imagination and intuition, through acts of human freedom and aesthetic expressions, and through cultural production. At the same time, I also leaned on Muslim chaura stories of creation, notions of human reason and freedom, thwarted histories of coexistence with Hindus, and understandings of nature as both God-given and subscribing to laws, espousing an easygoing naturalism that enabled a line of connection between them and the early European naturalists.

I hope that through this ethnography I have shown chaura attraction—which is also a compulsion—toward this way of life. I have desisted from calling it a form of life both because it isn't long-lived at any one site, as the river moves and impacts different people at different sites, and because the chauras are somewhat thrown together rather than being a cohesive, homogeneous group. I end with a short exposition that aims to capture the quality of this attraction/compulsion. Thus, when I say that this way of life is under threat from forces rife in the world, say those of climate change, I mean that the structure and substance of this desire as an aspect of nature is at risk as much as a people and a landscape.

Chauras view themselves as Adamites, descendants of the prophet Adam, through the close association of their livelihood as farmers with the Muslim Bangladeshi perspective of Adam as being created out of clay and working on soil (Thorp 1978). Here they draw on the Quranic verse on the creation of Adam that says: *He it is who created you from clay, and then decreed a stated term (for you).* It is likely the image of a hand shaping a figure out of earth and then breathing life into him that informs the popular imagination on the emergence of humans. Murata and Chittick (1998) provide a further reading of God's creation of Adam from earth that provides a backstory on racial diversity among humans and how they are never too far from either the Creator or earth. The companions of the Prophet relate a lively story of how earth and its many hues came to be related to humans in their racial diversity. They tell that when God wanted to create Adam and deemed that he would do so out of soil from earth, he sent down the angel Seraphiel to collect it. The earth protested. It did not want to give anything of itself. It begged and pleaded to be left intact. Seraphiel returned empty-handed. The angel Michael tried next, but he too was unsuccessful. Gabriel's trip was similarly futile. Finally, the angel Azrael was sent by God with special instructions. Azrael requested earth to give some soil in loan for a short period of time, promising that he personally would return the soil. At earth's agreement, Azrael gathered soil from different parts of the earth, red, white, black, and yellow. Thus these came to represent the different races of man and Azrael came to be the angel of death as he took human life and returned the body to the earth in accordance with his promise to earth. There is an interesting line of connection between this story of how the human body is on loan from the earth and the Mohini Chowdhury verse quoted earlier: *The earth/world desires me, don't keep me tied*. The singer and the earth/world exert a pull on each other. If the singer is not allowed to go, be free to roam, then the earth/world will come to him/her, be it in a dream or in the form of death. This is receptivity combined with desire.

Introduction

1. I sometimes use the expressions "form of life" or "mode of existence" when I wish to indicate there were some shared agreements among the chauras as to what constituted their lives.

2. Rising ocean waters and salinization of land were not the only ways in which chaura lives were threatened by climate change. Cyclone surges due to increased activity in the oceans were another. The impact of global warming on the circulation of winds necessary for the monsoons in the region was another major way in which the area might see climate change. Increased glacier melt and erratic or intensified rainfall could also be contributing to increased flooding. Climate change manifests in greater variability and uncertainty, rather than predictable, albeit catastrophic change.

3. The chauras' capacity for being different to themselves—in other words, espousing different sets of agreements on the nature of the social—recalls "Eskimo" society, of which Marcel Mauss ([1950] 1979) writes that it was not necessarily shaped by the topography of the landscape, which was in between water and land in consistency, as claimed by anthrogeography, but was communal in the winters and individualist in the summers, with the switch from one structural form to the other correlated to the seasons without being entirely determined by them.

Chapter 1. Moving Lands in the Skein of Property and Kin Relations

1. Marilyn Strathern (2009) claims that in certain cases, land as property may be productively thought of as not just being owned by people but also simultaneously owning people, who are called on to work on it. In other words, land is productive, and the people are one of its products. Through her extension of the concept of intellectual property to this situation, we are able to see how a people can feel entitled and duty bound to the land. Strathern's articulation somewhat describes the two-way relation between the chauras and the chars that I am charting, in which the chars call on the chauras to settle and cultivate them, and the chauras feel duty bound to realize the chars' productivity. Thus, there is a consonance between the two, but this consonance is not claimed on the basis of originary inhabitation. In other words, the chauras do not feel entitled to char land just because they settled it and have a long, ancestral history of interaction or property rights over it, but rather on the grounds that they are the only ones to answer its call as it keeps reemerging. I go further in arguing that the chauras respond to the tug of the land on them even when the land is not present, and that this tug of land, operating as an intangible property, a submerged potential, is what keeps chauras in their environment and working to ensure that land reemerges. This chapter describes two ways, physical and legal, in which this labor is undertaken; other ways will be presented in the later chapters.

2. For us to understand where someone like Strathern, with her notion of land laying claims on its products, including people, fits within this history of property, we have only to reprise the anthropological analysis of property by thinkers such as Durkheim (1983) and Gluckman (1965), who specified that the possessing person, the one who claims property or rights to property, derives from an originary community or status relation that asserts its hold on the person as either a member of the community or a status holder. It is only the person's relation to the community or position within a set of relations that authorizes the person to set aside any thing for their exclusive use. In other words, property is not exclusively a relation between a person and a thing but rather a relation between the person and other persons, whether they be other members of the community or other status holders. Property mediates relations. Furthermore, the derivation of the person's right to the property from an originary community or status holding means that that community or status lays a prior claim on that person, akin to land laying a prior claim on people as in the case spoken of by Strathern. I thank Sruti Chaganti for her insightful essay "On Property and Personhood" (2015) that helped me to delineate the anthropological position on property and a possible line of connection to Strathern.

3. Interestingly, it is largely land that materializes property rights and relations, in both the chaura context and many others. What is it about land that allows it to serve as the condition of possibility for property? One ready explanation is that given the antiquity and generality of agriculture, land is an image ready at hand. The other more anthropologically interesting speculation is that if our understanding of the human comes from our notion of ourselves as a practical species, pace Marx, shaping the world that the species finds around itself, then land is a materialization of the world.

4. I take owners to be those whose names were on the record of rights and title deeds, whereas lands might be under adverse possession, with land laws recognizing the length of possession as significant in determining actual or meaningful possession (Sirdar 1999). This was an artifact of British colonial law that grew to be in favor of raiyats or those possessing the right to hold or cultivate land. Or, in the case of char lands in which owners had sold their lands and the land had changed hands subsequently but without recourse to the formal, administrative means to record these changes, the new owner was only recognizable as possessor.

5. Daniel Miller (2007) makes a similar argument. Taking the example of Britain, he cautions that while in keeping with the analysis of new kinship "relationship" is certainly the primary way by which people make kin, in certain domains of life, such as in the practice of inheritance, older kinship norms, normative and legally airtight, seem to preside. Here he finds "an often almost desperate desire to repudiate experience in order to remain consistent with the imperatives of that formal order" (Miller 2007, 538).

6. While the writings on Bengali kinship do not fall under the rubric of new kinship studies, they are influenced by David Schneider (1984), specifically his intervention into old kinship, as a result of which scholars began to ply kinship terms not as part of some underlying structure to be uncovered but as symbolic of social relations. Transposing these thoughts to Bengali Muslim kinship, which is largely Bengali kinship (Fruzzetti and Östör 1976; Inden and Nicholas 2005), we might say that the importance that is given to blood (*rokto*) in such societies does not indicate the primacy of descent in kinship structures but rather that rokto is an important symbol by which to understand who belongs to whom and how. So while a child is a blood relative with the father's family, the child is also a blood relative with the mother's family. Rokto is a term to indicate this relationship (shoreek), but it also exists alongside kutum, those made relatives through marriage, with whom there are no blood relations but with whom relations may still be strong, as indicated by the quality of nearness or being *nikot* (Fruzzetti and Östör 1976). Generally, Bengali society, including Muslims, is patriarchal, patrilineal, and patrilocal. Marriage, particularly among Muslims, is exogamous, with

the wife relocating to the husband's home after marriage and taking on his gusti, or clan, which is one of the branches of a larger bongsho, or patrilineage (Arens and van Beurden 1977). While she retains her blood relations with her natal family, she also becomes shoreek, or included within the category of blood relations in her husband's lineage, indicating that this is not descent in the classical understanding of it—that is, kinship recapitulating biological reproduction. Furthermore, kutum include relatives far in excess of the usual in-laws and are better characterized by relations of giving and sharing rather than by marriage alone (Inden and Nicholas 2005). Consequently, drawing on chaura usage, I use the frequently used terms shoreek and kutum, instead of the usual concepts of descent and alliance, when speaking of chaura kin relations. A. H. M. Zehadul Karim also notes the large prevalence of fictive kin among rural Muslims in Bangladesh in which the term dhormo is affixed to usual kin relationships such as baap (father), ma (mother), bhai (brother), and bon (sister). Literally dhormo means "religious," and its use signals a relationship between two otherwise unrelated persons. It was usually an asymmetrical relationship between a powerful person and a less powerful person by which the latter gains some assistance, protection, and perhaps prestige (Karim 1990, 80–81). It merits mention because dhormo relations were in wide evidence in the chars.

7. An introduction to Shohidul is warranted as he appears frequently in my book. He began as a guide lent to me by MMS and then became my full-time research assistant. I initially relied on him to introduce me to people, transport me on his motorbike, and keep me company in male-only spaces and events, but he soon became a friend and an invaluable interlocutor. He hailed from the village of Bishtipur, or Village of Rain, in Sirajganj that went underwater in the 1980s. His large family had been prominent in Bishtipur, with epic stories attached to their names. After the breakage, Shohidul's father, a village teacher, moved to Nagarpur in nearby Tangail and set up his household there. Through a tremendous stratagem, his father acquired a good deal of contiguous property in this area that led to his family becoming cultivators on a large scale. They were also widely respected for their education and piety. Shohidul's family background and NGO experience were invaluable in giving me privileged access to sensitive topics, such as land-related information. In addition, Shohidul was a talented storyteller with a prodigious memory, and he quickly acquired a strong anthropological sensibility.

8. Here is folklorist Saymon Zakaria's description of the dance of lathi khela: "Carrying colored sticks in their hands, they come down to the field. They start yelling out threats ... and sometimes engage in some mock aggressive comedic banter. Then they begin to circle the whole field. As the games start, the group splits into two to fight each other. They attack each other with their sticks, shouting out warnings, 'Don't

dare!'... As the fight goes on, a musician will try to interrupt.... A fighter replies, 'What?' This banter goes on and on for quite some time... verging into philosophical speculation. For instance, a fighter might say, 'There are two parts of me, one woman, one man. Right now the woman is fighting'.... *Lathi Khela* is skilled theater" (Zakaria 2011, 184–85).

9. See Jenneke Arens's *Women, Land and Power in Bangladesh: Jhagrapur Revisited* (2011) for its description of how khas land management toward redistributing it among the landless poor has eluded effective government application.

10. *Degree* most likely referred to legal decree.

11. Shalish has a long history in the context of Bengal. The panchayat (literally, "the council of five") had mediated civil and criminal disputes within the context of villages since medieval times. The British replaced these with village courts starting in 1919. These courts were taxed to mediate settlements and not try cases, that being the jurisdiction of the legal courts (Siddiqui 2005). The village courts have also been mobilized by the Bangladesh state and have been much critiqued for the courts' capture by local elites and reconstituted through the reform efforts of NGOs and international donors (Berger 2017). However, I would argue that the courts elude full capture by the state, external forces, and internal powers, and thereby retain their reputation and normative hold on rural life.

12. Almost all studies of changing rural structures and economies in Bangladesh note the rise of the value of agricultural labor as a result of the liberalization of agriculture over the course of the 1970s and 1980s. This liberalization consisted in the reduced cost of irrigation equipment, privatization of fertilizer export and sales, and the production of high-yield varieties of rice (R. Ahmed, Haggblade, and Chowdhury 2000; Westergaard and Hossain 2005). It ushered in the green revolution in Bangladesh (Arens 2011) and was part of a larger set of structural adjustments imposed on the country as a consequence of loans taken by Bangladesh at concessionary interest rates (A. R. Khan 2001).

13. It is not only the case that Hindus left for India during the 1947 partition or due to subsequent Pakistani persecution during the India-Pakistan War of 1965. Hindus continued to leave for India well after the formation of Bangladesh in 1971 (Guhathakurta 2012).

14. In her classic study of rural society in Bangladesh shortly after it gained its independence in 1971, Kirsten Westergaard notes that the abolition of the zamindari system in the early 1950s eliminated the zamindars, but without eliminating rent interest or redistributing land, the East Bengal State Acquisition and Tenancy Act in effect left the entire structure intact to be run by *talukdars*, earlier tasked by the zamindars to collect rent, and *jotedars*, who were effectively wealthy raiyats or peasants. The continuation of an older way of being, with only some statutory changes,

is characterized by the preponderance of moneylending, states of indebtedness, and sharecropping (Westergaard 1985). This would explain early writings on village societies in Bangladesh that characterized its villages as still dominated by the culture of zamindars (see Zaidi 1970; A. Islam 1974).

15. Inheritance law in Bangladesh, notably within the Muslim Personal Law Application Act (1937) and the Muslim Family Laws Ordinance (1961), follows the prescriptions of Sharia, or Islamic law, which states that all property be distributed to legal descendants according to a set formula, with wives of deceased heads of household receiving one-fourth of their husband's holdings if there are no children or one-eighth if there are children, with the rest being distributed equally among sons, with each daughter getting half the share given to a son. Other males and females are also entitled, particularly if the deceased dies without leaving a male issue. The deceased may also stipulate in his will if he wishes to leave any portion of his property to nonentitled persons. However, Ansef did not inherit the land from his uncle Thandu. He gained the land through a resolution mediated in shalish. Whether this resolution was binding for all time and how it was made so binding was not something I could elicit from Ansef, but it would explain his desire to ensure his hold over the land through marriage relations.

16. In Edmund Leach's magisterial study of kinship and property relations in a village in Sri Lanka, Pul Eliya (1961), he claims that kinship is merely another way to speak about property relations. In his detailed case studies he shows how the normative kinship order did not in any way prescribe which fights over land emerged and how these conflicts were ultimately resolved. The many arrangements that were seen through these fights indicated that kinship was epiphenomenal. While I agree with Leach that the normative kinship order is not always predictive of property-related fights and their resolutions, I think he misses the opportunity to see how distant parts of a widespread kinship order were activated by these fights and how these activated parts may have made the fights as much about kinship as property rights, about bringing pockets of flexibility and negotiation within one's social reality. In fact, I would go so far as to claim that kinship and property are coextensive.

17. See Bryan Maddox (2001), in which he explores the different types of literacies, what he calls "subaltern literacies," present in economic practices in rural Bangladesh.

18. The predominant inheritance structure, outlined in note 15, is considered to be very schismatic, having led to a fragmentation of landholdings across Bangladesh (A. Rahman 1986). At present an average Bangladeshi has access to only 16 decimals of land (about .06 hectare), one of the smallest person-to-land ratios in the world, which effectively counts as landlessness. Within this context, inheritance does not carry

the weight and promise that it does within other capitalist societies (see Yanagisako 2015). As Atiur Rahman (1986) and Michael Harris (1989) note, inheritance is no guarantor for avoiding poverty and landlessness, with many of those people inheriting small plots of land, which are often not contiguous, effectively selling them off or using them as collateral for a loan and losing the land that way. These authors write that this fragmentation has worked to the benefit of currently landless people who are able to afford to buy the small plots of land that come into the market through the process described in this chapter and thereby consolidate contiguous holdings to move out of landlessness. There is no major concentration of land in the hands of farmers (although this is not the case for industrialists or even the state) as there is a ceiling of 33 bighas (approximately 20 acres) that a person can own. This may be because the state and the World Bank privilege small farm holdings as the most effective way to cultivate land in Bangladesh (M. H. Khan 2004).

19. It is useful to keep in mind the status of girls and women in chaura society. While the birth of girls weighed heavily on their parents because of dowry considerations, chauras by and large did not discriminate between their sons and daughters. They fed and provided for both as best they could, and it was the demonstration of intelligence that ensured the education of one child over another. By and large most boys and girls received only an elementary-level education, due as much to high demand for their labor and the itinerancy in their lives produced by floods and erosion as to budgetary constraints. While both girls and boys were expected to labor from early in their lives, there was considerable gender segregation, with girls rarely venturing out of their homes except to go to school and visit with relatives. This gave chaura girls a good reputation and made them desirable as wives. Parents practiced hypergamy to the extent possible because they wanted their daughters to leave the chars and live more comfortable lives on the mainland. They also tended to give their daughters in marriage much earlier than eighteen years of age, as demands for dowry were lower for younger girls. Char lives being considered difficult, this made it hard for males of marriage age to practice hypergamy. Most often they married within the extended family, lineage, the village, or the area.

20. Arens (2011) presents systematic findings that are close to my own in that she shows that more women in Bangladesh lay claims on their inheritance from their fathers in 2009 than in 1974 but delay those claims until after their parents are deceased. They also rely on intergenerational relations, such as the affection of an uncle for his sister's son, to facilitate such transfers. She also notes that most studies on women's access to land in Bangladesh focus exclusively on women's rights to their fathers' property but miss the fact that women also access land through rights to their husbands' property upon their husbands' deaths and through

purchasing land themselves. These are important dimensions to keep in mind in studying women and land in Bangladesh. At the end of her study, Arens is unable to conclude that landownership necessarily provides women more power and protection within society, but she is able to claim that such ownership is preferable to the elusive ideal of women's empowerment pursued by population control projects, microcredit, and social enterprise.

21. How is this extension of matter through mind philosophically thinkable? After all, for Descartes, matter was inert, if not dead. Kant, after Newton, saw matter as dynamic, held together through the balance of the forces of repulsion and attraction. This dynamism was a foundational insight for Schelling's *Naturphilosophie*. If matter was dynamic, then mind necessarily grew out of matter and retained matter's self-organizing properties. Products of the mind were material insofar as they shared the same origins, forces, and properties as matter. Mind was an extension of matter, lending itself to matter to render it intelligible, and to elaborate, extend, and realize matter's capacities. This was a companion process to the subject coming into consciousness within transcendental philosophy.

22. Within Schelling's transcendental philosophy, the "ought" had the nature of a will and was a demand placed by consciousness on oneself. It was the occasion for the self to come into consciousness of itself as object. I thank Andrew Brandel for drawing this to my attention.

23. See Sachiko Murata and William Chittick (1998) and S. Parvez Manzoor (2003) for a comprehensive introduction to the Islamic approach to creation and the divine in nature.

Chapter 2. History and Morality between Floods and Erosion

1. The outsize importance of floods in the Bangladeshi imaginary has been commented on. In their book *Floods in Bangladesh* (2006), Thomas Hofer and Bruno Messerli juxtapose the numbers killed and rendered homeless by floods, river erosion, and cyclones in Bangladesh for the past two centuries. They show that erosion made homeless the largest numbers of people, and cyclones killed the largest numbers of people, but that floods had commanded the most media attention, funds, and projects of the three. There is no definitive explanation for this disparity. It might be based on the spectacular nature, extensive spread, duration, and repetition of catastrophic floods that caught the international and national imagination, whereas, for instance, erosion tended to be more ongoing and localized. Certainly, floods had increased in frequency and intensity since the last century in South Asia, but so had river erosion and coastal cyclones, joined more recently by landslides in the hilly areas. In their introduction to *Water Resource Development in Bangladesh: Historical*

Documents (2010), Salim Rashid and Rezaur Rahman suggest that a reason for the historical focus on floods might be the long-standing concern with food security in East Bengal, then East Pakistan and now Bangladesh, which was considered to be seriously hampered by floods. Or else floods represented a type of hydrological problem that appeared amenable to engineering solutions (Brammer 2004), or they were another way to conduct geopolitics in the region as very often floods were accentuated by activities upriver located in other nation-states, such as China, Nepal, and India (Hofer and Messerli 2006; Ali 2010). Be that as it may, I take floods to be iconic of Bangladesh, both internally and externally.

2. The most historic of these events was West Pakistan's mishandling of the Bhola cyclone of 1970 in the Bay of Bengal of East Pakistan in which 500,000 people lost their lives to a sudden water surge (Nabil Ahmed 2013). The emergency response was so belated and relief so woefully inadequate that the West Pakistani military government was immediately assailed by international criticism. The event was marked by the benefit concert for Bangladesh organized by the musicians Ravi Shankar and George Harrison. And it precipitated the landslide victory for the East Pakistan–based Awami League party (Oldenburg 1985; Ali 2010).

3. Although scholars such as Tom Abel (1998) have tried to cast such social reproduction in terms of interaction between cultural and ecological dynamics, I do not think culture is as salient in this instance because the chauras did not have a long history of living with the river. The river has been moving along an east-west axis over the past two centuries. This movement has meant that the river draws new, untested people into it while leaving previous chauras on more stable land. In other words there was not an adequate buildup of experience, symbolization, and memory for intergenerational transmission before the river moved. I have also noticed that those who had moved away or had not experienced erosion in recent memory tended to reencounter chars as foreign and frightening, indicating how chars were always other to established Bengali culture and were quick to be othered once char dwellers were assimilated into mainland life. So, if char dwellers did what they had to do when chars eroded and their homes broke, it was in part because they were submitting to the suasion of the system, the system entraining them in the ways of the river, with mimicry and overhearing being other ways of acquiring knowledge (see N. Khan 2021b).

4. While this chapter is largely focused on localized floods and river erosion with attention to the underlying topography and the wider ecosystem of which they are a part (with topography and ecosystems being closely interrelated), the monsoons are part of the global atmosphere, referring to the seasonal reversal of winds and attendant patterns of rainfall, that helps produce the annual weather cycle in the tropical and

subtropical continents of Asia, Australia, and Africa. The driving planetary mechanisms of the monsoons are "1. *the differential heating of the land and ocean* and the resulting pressure gradient that drives the winds from high pressure to low pressure, and 2. *the swirl introduced to the winds* by the rotation of the earth" (Webster 1987, 8). Although there are other sources of rainfall in South Asia than the monsoons alone, including the western depressions of winter and the early summer thunderstorms known as the nor'westers, the summer rains from the monsoons are the most significant, particularly for the predominantly rain-fed agriculture in the area (H. E. Rashid 1991). While there are different triggers for floods in Bangladesh, monsoon rainfall is one of the strongest triggers for them (Hofer and Messerli 2006), with this rainfall also one of the relays of global climate change to the region (Brammer 2014). In other words, the greater frequency and intensity of floods in Bangladesh may be increasingly caused by monsoon rainfall pattern changes produced by climate change rather than by highland erosion in Nepal as has been long claimed (Hofer and Messerli 2006). So if floods are already national events, they may be on their way to becoming global ones. However, scientists caution prudence in analysis as the monsoons have always exhibited a great deal of variability, both seasonal and within seasons, with some years marked by strong monsoons (plenty of rain) and others by weak ones (drought-like conditions) and are also correlated to the Southern Oscillation that produces El Nino (T. Islam and Neelim 2010; Akter et al. 2016).

5. Hofer and Messerli (2006, 30) provide the following useful typology of floods in Bangladesh:

   - normal monsoon floods of major rivers, with the water level showing a slow rise and fall with pulsations;
   - flash floods, where hydrographs rise and fall sharply;
   - floods owing to excessive rainfall; and
   - tidal floods as a result of cyclones and storm surges.

6. Twenty-four decimals was the amount that was normatively pegged as the requisite for an ordinary chaura household with four inward-looking building structures with a central courtyard, although I heard of and saw households on much less, as low as five decimals.

7. The moral imperative for landowners to grant refuge, whether by selling small plots of land to those who sought it or by giving it in lease, in exchange for a deposit, or in some cases giving it free of charge, drew from the collective understanding that one day one would have to ask for refuge in turn. Those refusing to give land today were effectively denying the possibility that they might one day be reduced to asking for refuge. This response could be coming from a newfound confidence, some called it arrogance, that land on this char was now more or less permanent on account of the Jamuna Multipurpose Bridge (built in 1998) that had

apparently reduced the intensity of this channel of the river, although it intensified instability elsewhere. Or it could be that land on the char was now being used for cultivation of irri, a high-yield rice crop, which brought in surplus rice and consequently wealth for the landowners. This perhaps gave them the ability to disregard this code of conduct by buying land in the mainland, farther north such as in Rajshahi, where land was still relatively cheap.

8. The chauras of my acquaintance disputed Nojrul Chairman's account, saying that floods were a mere temporary inconvenience, with the land underneath becoming available again after floodwaters receded, whereas erosion meant land lost indefinitely. The loss of land wasn't just temporary harm; it was prolonged suffering (see also Mamun and Amin 1999; Abrar and Azad 2004).

9. As Hugh Brammer (2004) and Shapan Adnan (2008) have noted, the Flood Action Plan undertaken by the government of Ershad, who ruled from 1982 to 1990, irked the national intelligentsia and populace on account of its authoritarian approach to water management without the involvement of people. While Adnan and others further focus on the FAP as a possibly significant source of environmental problems for Bangladesh through its engineering approach to flood control and instead advocate flood-proofing (raising houses on plinths, creating flood shelters), Brammer discounts many of their concerns, showing how the FAP was never entirely engineering based and that it was crafted in response to people's desires not to be quite so vulnerable to floods. These debates have been ongoing even after the FAP expired, as they are expressive of different approaches to living on an active delta.

10. Kirsten Westergaard (1985) gives a different account of the shortage of food grains and the ensuing famine in 1974. She claims that the shortage was not caused by the floods, as has been usually claimed, but rather by the hoarding and storage of grain by wealthy peasants. The Mujib government played up the floods to be able to ensure foreign assistance to buy it time. However, in reality, the situation of famine was exacerbated by the economy being on the brink of collapse. Merchants who wanted the state to commit to private enterprise and were supported by members of the bureaucracy, foreign aid agencies, and foreign investors in their bid had the state under pressure. In other words, the 1974 famine was the product of the tussle between the state's hitherto weak commitment to socialism and self-sufficiency and its elite citizens' commitment to liberalization. This early opening to capitalism was later upheld by Presidents Zia and Ershad throughout the 1980s and made Bangladesh into the model neoliberal capitalist state (see Jahan 2001).

11. In *Drowned and Dammed* (2006), Rohan D'Souza's historical analysis shows how the region's landscape has been transformed from flood adaptability to flood vulnerability. In other words, where previously

humans and nonhuman others lived with some measure of knowledge of water's pathways, now the very same constituents experience the upsurge of water as a disorder of nature.

Chapter 3. Elections on Sandbars and the Remembered Village

1. Similar claims have been made on behalf of previous elections in Bangladesh. See Ahmed Shafiqul Huque and Muhammad Hakim (1993).

2. Latif Biswas was the same minister I had seen set off to provide emergency funds to flood victims amid a throng of people, as described in chapter 2.

3. Nojrul Chairman was the char-based politician who made the distinction between erosion and floods in terms of khoti (damage) and koshto (suffering) (see chapter 2).

4. Qadir was also crucial to the flourishing market in land underwater, particularly in searching out and bringing forth possible candidates who might prove to be original owners of abandoned property with whom to transact (see chapter 1). See A. K. M. Aminul Islam's *A Bangladesh Village: Conflict and Cohesion* (1974) for a study of the role of brokers between village-level politics and the nation-state in the early years of Bangladesh. University students were very important as mediators between the state and rural population in the early years of East Pakistan's formation in 1947.

5. See A. H. M. Zehadul Karim (1990) and Kamal Siddiqui (2005) for detailed accounts of the changing structure of political administration in Bangladesh from colonial times to the most recent past.

6. The village in East Pakistan/Bangladesh has a long history of study. However, in contrast to colonial times, when village studies were dictated by an interest in administering Indians, and early postcolonial village studies focused on India and its villages as objects of sociological curiosity, the study of villages in East Pakistan/Bangladesh has almost always been from the perspective of national development, given Bangladesh's particular openness to foreign aid, global capital, and neoliberal policies. In other words, villages have been cast as backward whose integration into the nation-state and espousal of nationalist ideology have been suspect, making them the focus of concerted efforts at modernization (e.g., Zaidi 1970; A. Islam 1974). Village development started as early as 1954 through a program known as Village Agricultural and Industrial Development, referred to as V-AID (Zaidi 1970). Alternatively, villages have been cast as fragile entities inadvertently caught in the swirl of external forces (e.g., Arens and van Beurden 1977; Westergaard 1985; S. White 1992). Over the decades, certain studies of villages with pseudonyms have come to occupy the status of classics to which scholars have responded in every decade. Boringram, or the Red Soil Village, made famous by Kirsten Westergaard (1985), was restudied by her and Abul Hossain (2005), show-

ing Bangladesh's politics and economic policies to have produced village-wide prosperity. Jhagrapur, or the Quarrelsome Village, first studied by Jenneke Arens and Jos van Beurden (1977), was restudied by Arens in the mid-1990s to show how the situation of women had gotten more precarious with the loss of traditional household-based livelihoods, the intrusion of population control measures, and the lingering inability to access resources by widows and divorcees. The study of Daripalla, the village made famous by John Thorp (1978), was perhaps the only one of such studies that did not make religion epiphenomenal to politics or economy. Thorp explored the Islamic cosmology that informed the identity of Muslim Bengali farmers and showed how Quranic narratives on the *umma* and traditional village forms, such as the shomaj, supported each other. Shapan Adnan (1997) restudied Daripalla in the 1990s to show the changing nature of shomaj. Village studies have been on the decline since the first decade of the twenty-first century, perhaps because the earlier concern over the Bangladesh village's inscrutability and impenetrability has been mitigated through extended rural upliftment programs and repeated studies, leaving villages, now immensely penetrable, to once again animate imaginations of autonomy and self-government.

Chapter 4. Decay of the River and of Memory

1. It is important to note here that Schelling is not claiming that it is nature in the human unconscious that drives humans to either good or evil. Rather, nature's presence in humans rounds out human acts, giving them a sense of completeness. So, in the instance of chauras in the throes of experiencing erosion, the expression of morality is still very much a human act but one in which the conjoining of erosion to self-opacity and to the compulsion to demand help or the compulsion to help each other suggests some aspect of the unconscious at work that gives chaura acts undertaken during erosion the feeling of appropriateness, even of necessity.

2. There is a long history of anthropologists studying absent presences. This work of course includes work by Stefania Pandolfo (1997) that shows that the Moroccan colonial past is far from the settled picture given to it by history but instead has a roiling quality and works its way through people's memories, dreams, and interrelations. Carlo Severi's study of the figure of the white man among statuettes used by Kuna shamans suggests that the figure is not symbolic revenge by the Kuna upon their colonizers. Rather, it represents the trace of a past of violence that has not yet entered into consciousness, but also refuses to fall into oblivion (Severi 2001). In James Siegel's haunting essay "The Hypnotist" (2011), he returns to his field site in Aceh in the aftermath of the 2004 tsunami to find the landscape filled with the missing, whose status as missing made it difficult to reconcile to their possible death. Instead, "the hypnotist

[an emergent menace to public order], installing death on the street, causes the evolution of the effects of the tsunami. It is not a question of the historical, singular loss of a person. Not like the woman who lost her daughters and whose memory of the tsunami focuses on that loss. It is rather a generalization of destruction, of loss, and of death. . . . It normalizes death as a property of the living" (115). In each of these instances we see the failure, or rather the limit points, of the representations of loss, but at no point do the authors abandon the concept of mediation in pursuit of some unmediated reality. Rather, the understanding is that unrepresentable forces will seep out at the edges of representation, that there is yet work to be done of staying the course of listening, attending, and analyzing.

3. Ralph Nicholas (2001) writes that Vaishnavism had found a stronger footing in rural Bengal than Shaivism by the late nineteenth century because of its structural parallels with Muslim notions of teacher-student relations and publicness.

4. As Rafiuddin Ahmed writes in *The Bengal Muslims* (1982), it was precisely the activity of itinerant Muslim scholars, trained in village and *mofassil* madrasas (small-town religious seminaries), that would lead to an Islamization of the Bengali countryside over the course of the late nineteenth century to the early twentieth, as the precursor to the partition of India of 1947.

Chapter 5. Death of Children and the Eruption of Myths

Chapter 5 was previously published as "Living Paradox in Riverine Bangladesh: Whiteheadian Perspectives on Ganga Devi and Khwaja Khijir" in *Anthropologica* 58, no. 2 (2016): 179–92.

1. It is noteworthy that a large part of Veena Das's work has provided a necessary corrective to anthropology's notion of itself as giving voice to marginalized people through capturing their speech and actions in ethnography or through amplifying their acts of rebellion or transgression against the normative order. If this were voice, then it is superficial at best, an act of charity on the part of the ethnographer or an interpretative overreach as such acts of transgression may not mount a serious challenge to a societal order and may even be sanctioned or overwritten by it. Voice as understood by Das is both more elusive and more demanding of the anthropologist, as it may communicate in nonverbal ways; it may be no more than a slight movement but may shake the very grounds of a given way of life. Voice may have to be stolen or staked and is vulnerable to the violence and skepticism of another. Anthropologists have to be aware that they are not always privy to acts of voicing and if by some chance they are, they have to be vigilant to this frailty of voice and the responsibility toward it (see Das 1989, 1991, 1996, 1998). I use voice, in this

sense of the word, as something toward which one's interlocutors feel compelled but that simultaneously constitutes a risk to their existence.

2. Cassirer, though not conventionally understood to be a romantic, likewise inherits a return to Kant deeply resonant with that of the early romantics, particularly in relation to myth. Each understands myth not merely as "allegorical illustrations of abstract ideas but as a particular form of cognitive appropriation of reality of its own which is different to, but no less valuable than, conceptual thought as it appears in the sciences and in philosophy" (Steinby 2009, 55). Schelling, for example, insisted on the importance of a philosophy of mythology (and not simply a comparative science of myths) as a primordial disclosure of God's revelation (see Wirth 2007; Schelling [1842] 2007). Cassirer's (1953) neo-Kantianism, on the other hand, drew him back to an emphasis on the singular event, figure, name, or instance that gives rise to an entire world.

3. One of Lévi-Strauss's great contributions to the study of mythology was to shift analysis away from comparison at a distance from ethnographic reality, insisting instead that material encounters undergirded structural transformations. "Structural analysis," he writes, must "meet one condition: it must never cut itself off from the facts" (1996, 189). In one of the most striking examples of this principle, Lévi-Strauss tracks mythological accounts of twins and doubling from Indo-European, Greco-Roman, and Amerindian contexts, revealing, on the level of myth, the traces of colonial encounter. Taking his cue from this multiregister approach, Raymond Jamous writes, quoting Charles Malamoud's work on Vedic myths: "United among themselves by kinship and marriage ties, the Greek gods are sufficiently individualized to be able to form a society; their histories fit with each other and there is in sum a history of gods, even though the time of the gods does not have the same structure as that of men. The gods of Brahmanism are indeterminate in number, varying according to the point of view taken. They have no real genealogy and their identity is too labile for it to be contained in any system of kinship. The myths about the individual gods are invaded by the vegetation of quasi-myths, accounts whose function is to justify the name of these gods or then to show how such and such groups of gods or the gods *en masse* acceded to the status whereby they become the addressees of such and such offerings" (1994, 336). This effort to follow the reverberations of an encounter across registers and its consequences across regions of life stands behind my efforts in this chapter. While chaura lives are far from these Vedic texts on rites, they share some elements on the lack of distinction about gods and prophets that grounds Cassirer's assertion that mythic thinking makes present a world from a particular point of view without its fullest elaboration and differentiation.

4. I am grateful to Swayam Bagaria and Andrew Brandel for drawing my attention to this story and for helping me to think through the

relationship between Khijir and Ganga. For retellings of the myth, see Diana Eck (1982), Anna King (2005), and Wendy Doniger (2009).

5. Annu Jalais (2010) notes the reanimation of Kali as the deity of choice among women prawn-seed collectors in southern Bangladesh.

6. It is noteworthy that later traditions will see Khijir as commensurate with Mina Nath. I am grateful to Projit Bihari Mukharji for this association between Khijir and Mina Nath. For an introduction to the Nath tradition, see James Mallinson (2011). Also see Projit Mukharji's "Lokman, Chholeman and Manik Pir: Multiple Frames of Institutionalising Islamic Medicine in Modern Bengal" (2011), in which he explores how Hindu figures, gods, and sacred icons take the shapes of *deos*, or demons, within present-day astrological and medicine manuals such as *Kitab-e-Chholemania*, which derives from late nineteenth-century efforts to produce a Bengali Islamic medicinal tradition.

7. The classic statement on Varuna is Georges Dumézil's *Mitra-Varuna: An Essay on Two Indo-European Representations of Sovereignty* (1988). Bhrigupati Singh (2015) utilizes this work effectively to show how state power in India operates through a twinned mode of capricious control and concern for the welfare of people.

8. In her study of the interrelation of the Vaisnava and the Sakta traditions within the context of ecstatic religion in Bengal, June McDaniel (1989) writes that the figure of the "baby-husband" is distinctive to the Sakta tradition (associated with Shiva) and in wide circulation.

9. She was the sister of Shukkur Member, whom I wrote about in chapter 1. She remembered when mothers-in-law would oil and braid the hair of their daughters-in-law, pulling a wire through each plait so that it would curve up, whereas now women's plaits simply hung down their backs in a nondescript fashion.

10. Hafeez Zaidi (1970) reports a saying in the village where he worked in which if a dreaming woman sees herself picking a fruit she will become pregnant. If she sees herself eat a fruit, she will remain childless. The implicit link here between eating something sweet and fleshy and childlessness casts women as cannibals.

11. Laura Marks, the author of *The Skin of the Film: Intercultural Cinema, Embodiment, and the Senses* (2000), has claimed that cinematography doesn't just produce the effect of the immediate; it instantiates it through embodiment. While Dill-Riaz sounded similar to Marks, interestingly he wasn't defensive about the representational quality of his filmmaking. He said readily that it took hundreds of hours of filmmaking to produce the hour-long documentary, and that it was the product of tremendous artifice. What he sought was a mode of picturing that grasped the multiple time frames that pressed on any possible moment, replete with presences, absences, and anticipations. He aimed to intimate such a

presence, to trigger a sense of it, rather than producing the presence as a whole. Other than Marks's conceptual commitment to immediacy, I am not certain I understand the difference between the two conceptions of the work of cinematography.

12. I thank Basab Mullik for the reference to this song.

13. John Thorp (1978, 1982) provides a valuable exposition on the theology and cosmogony of Bangladeshi Muslim farmers.

Abel, Tom. 1998. "Complex Adaptive Systems, Evolutionism, and Ecology within Anthropology: Interdisciplinary Research for Understanding Cultural and Ecological Dynamics." *Journal of Ecological Anthropology* 2 (1): 1–29.

Abrar, C. M., and S. N. Azad. 2004. *Coping with Displacement: Riverbank Erosion in North-West Bangladesh.* Dhaka: Refugee and Migratory Movements Research Unit.

Adnan, Shapan. 1997. "Class, Caste and Shamaj Relations among the Peasantry in Bangladesh: Mechanisms of Stability and Change in the Daripalla Villages, 1975–86." In *The Village in Asia Revisited,* edited by Jan Breman, Peter Kloos, and Ashwani Saith, 277–310. Oxford: Oxford University Press.

Adnan, Shapan. 2008. "Intellectual Critiques, People's Resistance and Inter-Riparian Contestations: Constraints to the Power of the State regarding Flood Control and Water Management in the Ganges-Brahmaputra-Meghna Delta of Bangladesh." In *Water, Sovereignty and Borders in Asia and Oceania,* edited by Devleena Ghosh, Heather Goodall, and Stephanie Hemelryk Donald, 104–24. London: Routledge.

Ahmed, Nabil. 2013. "Entangled Earth." *Third Text* 27 (1): 44–53.

Ahmed, Nizam. 2011. "Critical Elections and Democratic Consolidation: The 2008 Parliamentary Elections in Bangladesh." *Contemporary South Asia* 19 (2): 137–52.

Ahmed, Rafiuddin. 1982. *The Bengal Muslims, 1871–1906.* London: Oxford University Press.

Ahmed, Rafiuddin, ed. 2001. *Understanding the Bengal Muslims: Interpretative Essays.* New York: Oxford University Press.

Ahmed, Raisuddin, Steven Haggblade, and Tawfiq-e-Elahi Chowdhury, eds. 2000. *Out of the Shadow of Famine: Evolving Food Markets and Food Policy in Bangladesh*. Baltimore, MD: Johns Hopkins University Press.

Ahmed Syed Jamil. 2009. "Performing and Supplicating Manik Pir: Infrapolitics in the Domain of Popular Islam." *Drama Review* 53 (2): 51–76.

Ahmed, Syed Jamil. 2010. "Manik Pir Plays and a Subaltern Trickster: Grandiloquent Tales of Extra-scriptural Imagination." In *Folklore in Context: Essays in Honor of Shamsuzzaman Khan*, edited by Firoz Mahmud and Sharani Zaman, 113–32. Dhaka: University Press.

Ahmed-Siddiqi, Maulvi Mukhtar. 1916. *Sirajganjer Itihas*. Sirajganj, Bangladesh: Siddheswar Machine Press.

Ahuja, Amit, and Pradeep Chhibber. 2012. "Why the Poor Vote in India: 'If I Don't Vote, I Am Dead to the State.'" *Studies in Comparative International Development* 47:389–410.

Akter, Jahia, Maminul Haque Sarker, Ioana Popescu, and Dano Roelvink. 2016. "Evolution of the Bengal Delta and Its Prevailing Processes." *Journal of Coastal Research* 32 (5): 1212–26.

Al Faruque, Abdullah, and M. Hafijul Islam Khan. 2013. *Loss and Damage Associated with Climate Change: The Legal and Institutional Context in Bangladesh*. Dhaka: International Centre for Climate Change and Development.

Ali, S. Mahmud. 2010. *Understanding Bangladesh*. Oxford: Oxford University Press.

Arens, Jenneke. 2011. *Women, Land and Power in Bangladesh: Jhagrapur Revisited*. Dhaka: University Press.

Arens, Jenneke, and Jos van Beurden. 1977. *Jhagrapur: Poor Peasants and Women in a Village in Bangladesh*. New Delhi: Orient Longman.

Banerjee, Mukulika. 2007. "Sacred Elections." *Economic and Political Weekly* 42 (17): 1556–62.

Banerjee, Mukulika. 2011. "Elections as Communitas." *Social Research* 78 (1): 75–98.

Bangladesh Water Development Board. 2010. *Prediction of River Bank Erosion and Morphological Changes along the Jamuna, the Ganges and the Padma Rivers in 2010*. Dhaka: Center for Environmental and Geographic Information Services.

Baqee, Abdul. 1998. *Peopling in the Land of Allah Jaane*. Dhaka: University Press.

Barkat, Abul, Shafique uz Zaman, and Selim Raihan. 2001. *Political Economy of KHAS Land in Bangladesh*. Dhaka: Association for Land Reform and Development.

Bateson, Gregory. 1972. *Steps to an Ecology of Mind*. Chicago: University of Chicago Press.

Begum, Afroza. 2012. "Women's Participation in Union Parishads: A Quest for a Compassionate Legal Approach in Bangladesh from an International Perspective." *South Asia: Journal of South Asian Studies* 35 (3): 570–95.

Bennett, Jane. 2010a. *Vibrant Matter: A Political Ecology of Things*. Durham, NC: Duke University Press.

Bennett, Jane. 2010b. "A Vitalist Stopover on the Way to a New Materialism." In *New Materialisms: Ontology, Agency, and Politics*, edited by Diana Coole and Samantha Frost, 47–69. Durham, NC: Duke University Press.

Bennett, Jane. 2020. *Influx and Efflux: Writing Up with Walt Whitman*. Durham, NC: Duke University Press.

Berger, Tobias. 2017. *Global Norms and Local Courts Translating the Rule of Law in Bangladesh*. Oxford: Oxford University Press.

Bergson, Henri. 1911. *Creative Evolution*. New York: Camelot Press.

Bertocci, Peter J. 1970. "Elusive Villages: Social Structure and Community Organization in Rural East Pakistan." PhD diss., Michigan State University.

Bertocci, Peter J. 2002. "Islam and the Social Construction of the Bangladeshi Countryside." In *Understanding the Bengal Muslims: Interpretative Essays*, edited by Rafiuddin Ahmed, 71–85. New Delhi: Oxford University Press.

Best, James L., Philip J. Ashworth, Maminul H. Sarker, and Julie Roden. 2007. "The Brahmaputra-Jamuna River, Bangladesh." In *Large Rivers: Geomorphology and Management*, edited by Avijit Gupta, 374–93. New York: John Wiley and Sons.

Blanchet, Therese. 1984. *Meanings and Rituals of Birth in Rural Bangladesh: Women, Pollution and Marginality*. Dhaka: University Press.

Blum, Lauren S., Rasheda Khan, Adnan A. Hyder, Sabina Shahanaj, Shams El Arifeen, and Abdullah Baqui. 2009. "Childhood Drowning in Matlab, Bangladesh: An In-Depth Exploration of Community Perceptions and Practices." *Social Science and Medicine* 68:1720–27.

Boas, Franz. (1940) 1982. *Race, Language, and Culture*. Chicago: University of Chicago Press.

Brammer, Hugh. 2000. *Agroecological Aspects of Agricultural Research in Bangladesh*. Dhaka: University Press.

Brammer, Hugh. 2004. *Can Bangladesh Be Protected from Floods?* Dhaka: University Press.

Brammer, Hugh. 2014. *Climate Change, Sea-Level Rise and Development in Bangladesh*. Dhaka: University Press.

Brandel, Andrew. 2016. "The Art of Conviviality." *HAU: Journal of Ethnographic Theory* 6 (2): 323–43.

Breman, Jan, Peter Kloos, and Ashwani Saith. 1997. "Introduction." In *The Village in Asia Revisited*, edited by Jan Breman, Peter Kloos, and Ashwani Saith, 1–24. Oxford: Oxford University Press.

Brocklesby, Mary Ann, and Mary Hobley. 2003. "The Practice of Design: Developing the Chars Livelihoods Programme in Bangladesh." *Journal of International Development* 15:893–909.

Brown, Norman O. 1983. "The Apocalypse of Islam." *Social Text* 8:155–71.

Butler, Judith. 2005. *Giving an Account of Oneself*. New York: Fordham University Press.

Cashin, David G. 2010. "Nathist Folklore and Sufism in Bengal." In *Folklore in Context: Essays in Honor of Shamsuzzaman Khan*, edited by Firoz Mahmud and Sharani Zaman, 93–105. Dhaka: University Press.

Cassidy, Rebecca, and Molly Mullin, eds. 2007. *Where the Wild Things Are Now: Domestication Reconsidered*. New York: Berg.

Cassirer, Ernst. 1953. *Language and Myth*. Mineola, NY: Dover.

Castro, Eduardo Viveiros de. 2014. *Cannibal Metaphysics*. Edited and translated by Peter Skafish. Minneapolis: Univocal.

Chaganti, Sruti. 2015. "On Property and Personhood." Conceptual essay, Department of Anthropology, Johns Hopkins University.

Chatterji, Joya. 2002. *Bengal Divided: Hindu Communalism and Partition, 1932–1947*. New York: Cambridge University Press.

Chatterji, Joya. 2007. *The Spoils of Partition: Bengal and India, 1947–1967*. New York: Cambridge University Press.

Choudhury, Nurul H. 2001. *Peasant Radicalism of the Nineteenth Century: The Faraizi, Indigo and Pabna Movements*. Dhaka: Asiatic Society of Bangladesh.

Chowdhury, Afsan. 2009. "Hindus in a Polarized Political Environment: Bangladesh's Minority." In *Living on the Margins: Minorities in South Asia*, edited by Rita Manchanda, 35–49. Kathmandu: South Asia Forum for Human Rights.

Chowdhury, Jafar Ahmed. 2007. *Essays on Environment*. Dhaka: Botomul.

Chowdhury, Nusrat S. 2019. *Paradoxes of the Popular: Crowd Politics in Bangladesh*. Stanford, CA: Stanford University Press.

Church, Michael. 2006. "Bed Material Transport and the Morphology of Alluvial River Channels." *Annual Review of Earth and Planetary Sciences* 34:325–54.

Clayton, Martin, Rebecca Sager, and Udo Will. 2005. "In Time with the Music: The Concept of Entrainment and Its Significance for Ethnomusicology." *European Meetings in Ethnomusicology* 11, ESEM Counterpoint 1:1–82.

Coleman, James M. 1960. "Brahmaputra River: Channel Processes and Sedimentation." *Sedimentary Geology* 3 (2–3): 129–239.

Collingwood, R. G. 1960. *The Idea of Nature*. New York: Oxford University Press.

Cotton, James Sutherland, Sir Richard Burn, and Sir William Stevenson Meyer. 1908. *The Imperial Gazetteer of India*. N.s., vol. 19, *Nayakanhatti to Parbhani*. Oxford: Clarendon Press.

Darian, Steven G. 1978. *Ganges in Myth and History*. Honolulu: University of Hawaii Press.

Das, Veena. 1989. "Voices of Children." *Daedalus* 118 (4): 263–94.

Das, Veena. 1991. "Composition of the Personal Voice: Violence and Migration." *Studies in History* 7 (1): 65–77.

Das, Veena. 1995. "Voice as Birth of Culture." *Ethnos* 60 (3–4): 159–79.

Das, Veena. 1996. "Language and Body: Transactions in the Construction of Pain." *Daedalus* 125 (1): 67–91.

Das, Veena. 1998. "Wittgenstein and Anthropology." *Annual Review of Anthropology* 27 (1): 171–95.

Das, Veena. 2006. *Life and Words: Violence and the Descent into the Ordinary*. Berkeley: University of California Press.

Das, Veena. 2014. "Cohabiting an Interreligious Milieu: Reflections on Religious Diversity." In A Companion to an Anthropology of Religion, edited by Janice Boddy and Michael Lambek, 69–84. New York: John Wiley and Sons.

Das, Veena. 2020. "Concepts Crisscrossing: Anthropology and Knowledge Making." In Textures of the Ordinary: Doing Anthropology after Wittgenstein, 275–306. New York: Fordham University Press.

Descola, Philippe. 2013a. "Beyond Nature and Culture: Forms of Attachment." Translated by Janet Lloyd. HAU: Journal of Ethnographic Theory 2 (1): 447–71.

Descola, Philippe. 2013b. "Beyond Nature and Culture: The Traffic of Souls." Translated by Janet Lloyd. HAU: Journal of Ethnographic Theory 2 (1): 473–500.

De Wilde, Koen, ed. 2011. Moving Coastlines: Emergence and Use of Land in the Ganges-Brahmaputra-Meghna Estuary. Dhaka: University Press.

Doniger, Wendy. 2009. The Hindus: An Alternative History. London: Penguin Books.

D'Souza, Rohan. 2006. Drowned and Dammed: Colonial Capitalism and Flood Control in Eastern India. New York: Oxford University Press.

Dumézil, Georges. 1988. Mitra-Varuna: An Essay on Two Indo-European Representations of Sovereignty. New York: Zone Books.

Durkheim, Émile. 1983. Durkheim and the Law. Edited by S. Lukes and A. Scull. New York: St. Martin's Press.

Eaton, Richard M. 1993. The Rise of Islam and the Bengal Frontier, 1204–1760. Berkeley: University of California Press.

Eck, Diana L. 1982. "Ganga: The Goddess Ganges in Hindu Sacred Geography." In The Divine Consort: Radha and the Goddesses of India, edited by John Stratton Hawley and Donna Marie Wulff, 166–83. Boston: Beacon Press.

EGIS (Environment and GIS Support Project for Water Sector Planning). 2000. Riverine Chars in Bangladesh: Environmental Dynamics and Management Issues. Dhaka: University Press.

Engels, Friedrich. (1888) 2010. The Origin of the Family, Private Property and the State. London: Penguin Books.

Feldman, Shelley, and Charles Geisler. 2012. "Land Expropriation and Displacement in Bangladesh." Journal of Peasant Studies 39 (3–4): 971–93.

Fenves, Peter. 2010. The Messianic Reduction: Walter Benjamin and the Shape of Time. Palo Alto, CA: Stanford University Press.

Franklin, Sarah, and Susan McKinnon, eds. 2001. Relative Values: Reconfiguring Kinship Studies. Durham, NC: Duke University Press.

Fruzzetti, Lina, and Àkos Östör. 1976. "Seed and Earth: A Cultural Analysis of Kinship in a Bengali Town." Contributions to Indian Sociology 10 (1): 97–132.

Gain, Animesh K., Heiko Apel, Fabrice K. Renaud, and Carlo Giupponi. 2013. "Thresholds of Hydrologic Flow Regime of a River and Investigation of Climate Change Impact—The Case of the Lower Brahmaputra River Basin." Climatic Change 120:463–75.

Gandhi, Mohandas. 1996. "Hind Swaraj" and Other Writings. Edited by Anthony J. Parel. New York: Cambridge University Press.

Gilbert, Marc Jason. 1990. "Mofussil Municipal Reform in Late Nineteenth Century Bengal: Nationalism and Development on Trial." *Journal of Third World Studies* 7 (1): 84–115.

Gilmartin, David. 2012. "Towards a Global History of Voting: Sovereignty, the Diffusion of Ideas, and the Enchanted Individual." *Religions* 3:407–23.

Gluckman, Max. 1965. *The Ideas in Barotse Jurisprudence*. New Haven, CT: Yale University Press.

Goethe, Johann Wolfgang von. (1790) 2009. *The Metamorphosis of Plants*. Cambridge, MA: MIT Press.

Goethe, Johann Wolfgang von. 1988a. "Excerpt from 'Studies for a Physiology of Plants' [A Schematic Fragment]." In *Scientific Studies*, edited and translated by Douglas Miller, 73–75. Vol. 12 of *Goethe: The Collected Works*. New York: Suhrkamp Verlag.

Goethe, Johann Wolfgang von. 1988b. "Judgement through Intuitive Perception." In *Scientific Studies*, edited and translated by Douglas Miller, 31–32. Vol. 12 of *Goethe: The Collected Works*. New York: Suhrkamp.

Graburn, Nelson, ed. 1971. *Readings in Kinship and Social Structure*. New York: Harper and Row.

Guha, Ranajit. 1996. *A Rule of Property for Bengal: An Essay on the Idea of Permanent Settlement*. Durham, NC: Duke University Press.

Guhathakurta, Meghna. 2012. "Amidst the Winds of Change: The Hindu Minority in Bangladesh." *South Asian History and Culture* 3 (2): 288–301.

Guyer, Jane. 1996. "Traditions of Invention in Equatorial Africa." *African Studies Review* 39 (3): 1–28.

Hadot, Pierre. 2006. *The Veil of Isis: An Essay on the History of the Idea of Nature*. Cambridge, MA: Belknap Press of Harvard University Press.

Hann, C. M. 1998. "Introduction: The Embeddedness of Property." In *Property Relations: Renewing the Anthropological Tradition*, edited by C. M. Hann, 1–47. New York: Cambridge University Press.

Haque, Chowdhury E. 1988. "Adjustments to River Bank Erosion Hazard in the Jamuna Floodplain, Bangladesh." *Human Ecology* 16 (4): 421–37.

Haraway, Donna. 2003. *The Companion Species Manifesto: Dogs, People, and Significant Otherness*. Chicago: Prickly Paradigm Press.

Harris, Michael S. 1989. "Land, Inheritance, and Economic Mobility: An Example from Bangladesh." *Urban Anthropology and Studies of Cultural Systems and World Economic Development* 18 (3–4): 329–46.

Harrison, Simon. 2004. "Forgetful and Memorious Landscapes." *Social Anthropology* 12 (2): 135–51.

Hasan, Samiul. 1992. "Upazila Development Planning in Bangladesh: Problems of Resource Mobilization." *Asian Survey* 32 (9): 802–14.

Hatley, Shaman. 2007. "Mapping the Esoteric Body in the Islamic Yoga of Bengal." *History of Religions* 46 (4): 351–68.

Hill, Christopher V. 1997. *River of Sorrow: Environment and Social Control in Riparian North India, 1770–1994*. Ann Arbor, MI: Association for Asian Studies.

Hofer, Thomas, and Bruno Messerli. 2006. *Floods in Bangladesh: History, Dynamics and Rethinking the Role of the Himalayas*. New York: United Nations University Press.

Hoque, Muhammad Enamul. 1995. *Muhammad Enamul Hoque Rachanabali.* Vol. 4, *Complete Works of Muhammad Enamul Hoque*. Edited by Monsur Musa. Dhaka: Bangla Academy.

Hossain, Naomi. 2017. *The Aid Lab: Understanding Bangladesh's Unexpected Success*. London: Oxford University Press.

Hunter, W. W. 1886. *The Imperial Gazetteer of India*. O.s., vol. 10. Multan-Palhalli: Trubner.

Huque, Ahmed Shafiqul, and Muhammad A. Hakim. 1993. "Elections in Bangladesh: Tools of Legitimacy." *Asian Affairs* 19 (4): 248–61.

Inden, Ronald B., and Ralph W. Nicholas. 2005. *Kinship in Bengali Culture*. 2nd ed. New Delhi: Chronicle Books.

Indra, Doreen. 2000. "Not Just Dis-placed and Poor: How Environmentally Forced Migrants in Rural Bangladesh Recreate Space and Place under Trying Conditions." In *Rethinking Refuge and Displacement: Selected Papers of Refugees and Immigrants*, vol. 3, edited by Elzbieta M. Gozdziak and Dianna Shandy, 163–91. Washington, DC: American Anthropological Association.

Iqbal, Iftekhar. 2010. *The Bengal Delta: Ecology, State, and Social Change, 1840–1943*. New York: Palgrave Macmillan.

Iqbal, Muhammad. (1924) 2003. "Bang-e Dara" in *Poems of Iqbal*. Translated by V. G. Kiernan. Lahore: Iqbal Academy.

Islam, A. K. M. Aminul. 1974. *A Bangladesh Village, Conflict and Cohesion: An Anthropological Study of Politics*. Cambridge: Schenkman Publishing.

Islam, Nazrul M. 2010. *Braiding and Channel Morphodynamics of the Brahmaputra-Jamuna River*. Saarbrücken: Lambert Academic Publishing.

Islam, Tawhidul, and Ananta Neelim. 2010. *Climate Change in Bangladesh: A Closer Look into Temperature and Rainfall Data*. Dhaka: University Press.

Jabbar, Abdul. 1998. "Shonkhraj." In *Famous Ghost Stories*, edited by Leela Majumdar, 147–60. Calcutta: Kamini Press.

Jahan, Rounaq, ed. 2001. *Bangladesh: Promise and Performance*. Chicago: Zed Books.

Jalais, Annu. 2010. "Braving Crocodiles with Kali: Being a Prawn-Seed Collector and a Modern Woman in the 21st Century Sundarbans." *Socio-Legal Review* 6:1–23.

Jamous, Raymond. 1994. "Rites of Ancient India: Outlook for Comparative Anthropology." *Contributions to Indian Sociology* 28 (2): 323–52.

Jodhka, Surinder S. 2002. "Nation and Village: Images of Rural India in Gandhi, Nehru and Ambedkar." *Economic and Political Weekly* 37 (32): 3343–53.

Karim, A. H. M. Zehadul. 1990. *The Pattern of Rural Leadership in an Agrarian Society: A Case Study of the Changing Power Structure in Bangladesh*. New Delhi: Northern Book Centre.

Khan, Azizur Rahman. 2001. "Economic Development: From Independence to the End of the Millennium." In *Bangladesh: Promise and Performance*, edited by Rounaq Jahan, 247–66. Chicago: Zed Books.

Khan, Mohammed Ayub. (1909) 1967. *Friends Not Masters: A Political Autobiography*. Oxford: Oxford University Press.

Khan, Mushtaq Husain. 2000. "Class, Clientelism and Communal Politics in Contemporary Bangladesh." In *The Making of History: Essays Presented to Irfan Habib*, edited by K. N. Panikkar, T. J. Byres, and U. Patnaik, 1–45. New Delhi: Tulika.

Khan, Mushtaq Husain. 2004. "Power, Property Rights and the Issue of Land Reform: A General Case Illustrated with Reference to Bangladesh." *Journal of Agrarian Change* 1–2:73–106.

Khan, Naveeda. 2014. "The Death of Nature in the Era of Global Warming." In *Wording the World: Veena Das and Scenes of Inheritance*, edited by Roma Chatterji, 288–99. New York: Fordham University Press.

Khan, Naveeda. 2015. "Of What Does Self-Knowing Consist? Perspectives from Bangladesh and Pakistan." *Annual Review of Anthropology* 44:457–75.

Khan, Naveeda. 2021a. "Kant and Anthropology." In *Oxford Research Encyclopedia of Anthropology*. https://doi.org/10.1093/acrefore/9780190854584.013.311.

Khan, Naveeda. 2021b. "Marginal Lives and the Microsociology of Overhearing in the Jamuna Chars." *Ethnos*, June 3: 1–22.

King, Anna S. 2005. "The Ganga: Waters of Devotion." In *The Intimate Other: Love Divine in Indic Religions*, edited by Anna S. King and John Brockington, 153–93. New Delhi: Orient Longman.

Kirksey, S. Eben, and Stefan Helmreich. 2010. "The Emergence of Multispecies Ethnography." *Cultural Anthropology* 25 (4): 545–76.

Kohn, Eduardo. 2013. *How Forests Think: Toward an Anthropology beyond the Human*. Berkeley: University of California Press.

Kotalova, Jitka. 1993. *Belonging to Others: Cultural Construction of Womanhood among Muslims in a Village in Bangladesh*. Stockholm: Almquist and Wikseli.

Latour, Bruno. 2005. *Reassembling the Social: An Introduction to Actor-Network Theory*. New York: Oxford University Press.

Laugier, Sandra. 2015. "Voice as Form of Life and Life Form." In "Wittgenstein and Forms of Life," edited by Daniele Moyal-Sharrock and Piergiorgio Donatelli, 63–82. Special issue, *Nordic Wittgenstein Review* (October): 63–81.

Leach, Edmund Ronald. 1961. *Pul Eliya, a Village in Ceylon: A Study of Land Tenure and Kinship*. New York: Cambridge University Press.

Lévi-Strauss, Claude. 1962. *The Savage Mind*. Chicago: University of Chicago Press.

Lévi-Strauss, Claude. 1992. *Tristes Tropiques*. Translated by John Weightman and Doreen Weightman. London: Penguin Books.

Lévi-Strauss, Claude. 1996. *The Story of Lynx*. Translated by Catherine Tihanyi. Chicago: University of Chicago Press.

Lewis, David. 2011. *Bangladesh: Politics, Economy and Civil Society*. Cambridge: Cambridge University Press.

Lewis, David, and Abul Hossain. 2008. "A Tale of Three Villages: Power, Difference and Locality in Rural Bangladesh." *Journal of South Asian Development* 3:33–51.

MacPherson, Crawford Brough. 1962. *The Political Theory of Possessive Individualism: From Hobbes to Locke.* Oxford: Oxford University Press.

Maddox, Bryan. 2001. "Literacy and the Market: The Economic Uses of Literacy among the Peasantry in North-Western Bangladesh." In *Literacy and Development: Ethnographic Perspectives,* edited by Brian V. Street, 137–51. New York: Routledge.

Mahalanobis, P. C. 1927. *Report on Rainfall and Floods in North Bengal, 1870–1922.* Calcutta: Bengal Secretariat Book Department.

Mallinson, James. 2011. "Nath-Sampradaya." *Brill Encyclopedia of Hinduism* 3:407–28.

Mamun, Muhammad Z., and A. T. M. Nurul Amin. 1990. *Densification: A Strategic Plan to Mitigate Riverbank Erosion Disaster in Bangladesh.* Dhaka: University Press.

Manzoor, S. Parvez. 2003. "Nature and Culture: An Islamic Perspective." In *Nature across Cultures: Views of Nature and the Environment in Non-Western Cultures,* edited by Helaine Selin, 421–32. Dordrecht: Kluwer Academic Publishers.

Marks, Jonathan. 2017. *Human Biodiversity: Genes, Race, and History.* London: Routledge.

Marks, Laura. 2000. *The Skin of the Film: Intercultural Cinema, Embodiment, and the Senses.* Durham, NC: Duke University Press.

Marx, Karl. (1930) 1988. *Economic and Philosophical Manuscripts of 1844.* New York: Prometheus Books.

Massey, Doreen. 2006. "Landscape as a Provocation: Reflections on Moving Mountains." *Journal of Material Culture* 11 (1–2): 33–48.

Mauss, Marcel. (1950) 1979. *Seasonal Variations of the Eskimo.* New York: Routledge.

McDaniel, June. 1989. *The Madness of the Saints: Ecstatic Religion in Bengal.* Chicago: University of Chicago Press.

Mendelsohn, Oliver. 1993. "The Transformation of Authority in Rural India." *Modern Asian Studies* 27 (4): 805–42.

Menski, Werner F., and Tahmina Rahman. 1988. "Hindus and the Law in Bangladesh." *South Asia Research* 8 (2): 111–31.

Miller, Daniel. 2007. "What Is a Relationship? Is Kinship Negotiated Experience?" *Ethnos* 72 (4): 535–54.

Mirza, M. Monirul Qader, R. A. Warrick, and N. J. Ericksen. 2003. "The Implications of Climate Change on Floods of the Ganges, Brahmaputra and Meghna Rivers in Bangladesh." *Climatic Change* 57:287–318.

Mookherjee, Nayanika. 2007. "The 'Dead and Their Double Duties': Mourning, Melancholia, and the Martyred Intellectual Memorials in Bangladesh." *Space and Culture* 10 (2): 271–91.

Mookherjee, Nayanika. 2015. *The Spectral Wound: Sexual Violence, Public Memories, and the Bangladesh War of 1971.* Durham, NC: Duke University Press.

Morris, Rosalind C. 2017. "After de Brosses: Fetishism, Translation, Comparativism, Critique." In *The Return of Fetishism: Charles de Brosses and the Afterlives of an Idea*, edited by Rosalind C. Morris and Daniel H. Leonard, 133–36. Chicago: University of Chicago Press.

Mukharji, Projit. 2011. "Lokman, Chholeman and Manik Pir: Multiple Frames of Institutionalising Islamic Medicine in Modern Bengal." *Social History of Medicine* 24 (3): 720–38.

Mukherjee, Rila. 2008. "Putting the Rafts Out to Sea: Talking of 'Bera Bhashan' in Bengal." *Transforming Cultures eJournal* 3 (2): 124–44.

Murata, Sachiko, and William Chittick. 1998. *The Vision of Islam*. St. Paul, MN: Paragon Publishers.

Nassar, Dalia. 2013. "Intellectual Intuition and the Philosophy of Nature: An Examination of the Problem." In *Ubergange—diskursiv oder intuitiv? Essays zur Eckart Forsters Die 25 Jahre der Philosophie*, edited by Johannes Haag and Markus Wild, 235–58. Frankfurt: Vittorio Klostermann.

Nassar, Dalia. 2014. *The Romantic Absolute: Being and Knowing in Early German Romantic Philosophy, 1795–1804*. Chicago: University of Chicago Press.

Netton, Ian Richard. 1992. "Theophany as Paradox: Ibn al-'Arabi's Account of al-Khadir in His Fusus al-Hikam." *Journal of the Muhyiddin Ibn 'Arabi Society* 11: 11–22.

Nicholas, Ralph W. 2001. "Islam and Vaishnavism in the Environment of Rural Bengal." In *Understanding the Bengal Muslims: Interpretative Essays*, edited by Rafiuddin Ahmed, 52–71. New York: Oxford University Press.

Oldenburg, Philip. 1985. "'A Place Insufficiently Imagined': Language, Belief, and the Pakistan Crisis of 1971." *Journal of Asian Studies* 44 (4): 711–33.

Omar, Irfan A. 1993. "Khidr in the Islamic Tradition." *Muslim World* 83 (3–4): 279–94.

Omar, Irfan A. 2004. "Khizr-i Rah: The Pre-eminent Guide to Action in Muhammad Iqbal's Thought." *Islamic Studies* 43 (1): 39–50.

Omar, Irfan A. 2010. "Reflecting Divine Light: al-*Kihdr* as an Embodiment of God's Mercy (*rahma*)." In *Gotteserlebnis und Gotteslehre: Christliche und Islamiche Mystik im Orient*, edited by Martin Tamcke, 167–80. Wiesbaden: Harrassowitz.

Palmié, Stephan. 2018. "When Is a Thing? Transduction and Immediacy in Afro-Cuban Ritual; or, ANT in Matanzas, Cuba, Summer of 1948." *Comparative Studies in Society and History* 60 (4): 786–809.

Panday, Pranab Kumar. 2008. "Representation without Participation: Quotas for Women in Bangladesh." *International Political Science Review* 29 (4): 489–512.

Pandolfo, Stefania. 1997. *Impasse of the Angels: Scenes from a Moroccan Space of Memory*. Chicago: University of Chicago Press.

Penz, Peter, Jay Drydyk, and Pablo Bose. 2011. *Displacement by Development: Ethics, Rights and Responsibilities*. New York: Cambridge University Press.

Peterson, Garry D. 2002. "Contagious Disturbance, Ecological Memory, and the Emergence of Landscape Pattern." *Ecosystems* 5 (4): 329–38.

Povinelli, Elizabeth A. 2016. *Geontologies: A Requiem to Liberalism*. Durham, NC: Duke University Press.

Rahman, Atiur. 1986. "Differentiation of the Peasantry in Bangladesh: 1950s to 1980s." *Social Scientist* 14 (11–12): 68–95.

Rahman, Md. Mahbubar, and Willem van Schendel. 1997. "Gender and the Inheritance of Land: Living Law in Bangladesh." In *The Village from Asia Revisited*, edited by Jan Breman, Peter Kloos, and Ashwani Saith, 237–76. Oxford: Oxford University Press.

Rahman, Sabeel. 2006. "Development, Democracy and the NGO Sector." *Theory and Evidence from Bangladesh* 22 (4): 451–73.

Rappaport, Roy. (1984) 2000. *Pigs for the Ancestors: Ritual in the Ecology of a New Guinea People*. Long Grove, IL: Waveland Press.

Rashid, Haroun Er. 1991. *Geography of Bangladesh*. 2nd ed. Dhaka: University Press.

Rashid, Salim, and Rezaur Rahman, eds. 2010. *Water Resource Development in Bangladesh: Historical Documents*. Dhaka: University Press.

Riaz, Ali. 2014. "Bangladesh's Failed Election." *Journal of Democracy* 25 (2): 119–30.

Riaz, Ali. 2016. *Bangladesh: A Political History since Independence*. London: I. B. Tauris.

Roy, Asim. 1983. *The Islamic Syncretistic Tradition in Bengal*. Princeton, NJ: Princeton University Press.

Roy, Beth. 1996. *Some Trouble with Cows: Making Sense of Social Conflict*. Dhaka: University Press.

Sahlins, Marshall D. 1976. *The Use and Abuse of Biology: An Anthropological Critique of Sociobiology*. Ann Arbor: University of Michigan Press.

Sahlins, Marshall D. 2013. *What Kinship Is—And Is Not*. Chicago: University of Chicago Press.

Sarker, Maminul, Iffat Huque, and Mustafa Alam. 2003. "Rivers, Chars, and Char Dwellers of Bangladesh." *International Journal of River Basin Management* 1 (1): 61–80.

Sarker, Maminul, and Colin R. Thorne. 2006. "Morphological Response of the Brahmaputra–Padma–Lower Meghna River System to the Assam Earthquake of 1950." In *Braided Rivers*, edited by Greg Sambrook Smith, Jim Best, Charlie Bristow, and Geoff E. Petts, 289–311. Hoboken, NJ: Blackwell.

Sarker, Maminul, Colin R. Thorne, M. Nazneen Aktar, and Md. Ruknul Ferdous. 2014. "Morphodynamics of the Brahmaputra-Jamuna River, Bangladesh." *Geomorphology* 215:45–59.

Sayidur, Muhammad. 1991. *Bera Bhasan Utshob*. Dhaka: Bangla Academy.

Schelling, Friedrich Wilhelm Joseph von. (1797) 1988. *Ideas for a Philosophy of Nature*. New York: Cambridge University Press.

Schelling, Friedrich Wilhelm Joseph von. (1799) 2004. *First Outline of a System of the Philosophy of Nature*. Translated by Keith R. Peterson. Albany: State University of New York Press.

Schelling, Friedrich Wilhelm Joseph von. (1800) 1993. *System of Transcendental Idealism*. Translated by Peter Heath. Charlottesville: University Press of Virginia.

Schelling, Friedrich Wilhelm Joseph von. (1842) 2007. *Philosophical Investigations into the Essence of Human Freedom*. Albany: State University of New York Press.

Schelling, Friedrich Wilhelm Joseph von. (1842) 2012. *Historical-critical Introduction to the Philosophy of Mythology*. Albany: State University of New York Press.

Scheper-Hughes, Nancy. 1992. *Death without Weeping: The Violence of Everyday Life in Brazil*. Berkeley: University of California Press.

Schmuck-Widmann, Hanna. 2000. "'An Act of Allah': Religious Explanations for Floods in Bangladesh as Survival Strategy." *International Journal of Mass Emergencies and Disasters* 18 (1): 85–96.

Schmuck-Widmann, Hanna. 2001. *Facing the Jamuna River: Indigenous and Engineering Knowledge in Bangladesh*. Dhaka: Bangladesh Resource Centre for Indigenous Knowledge.

Schneider, David M. 1984. *A Critique of the Study of Kinship*. Ann Arbor: University of Michigan Press.

Severi, Carlo. 2001. "Cosmology, Crisis, and Paradox: On the White Spirit in the Kuna Shamanistic Tradition." In *Disturbing Remains: Memory, History, and Crisis in the Twentieth Century*, edited by Michael Roth and Charles Salas, 178–206. Los Angeles: Getty Research Institute.

Severi, Carlo. 2004. "Capturing Imagination: A Cognitive Approach to Cultural Complexity." *Journal of the Royal Anthropological Institute*, n.s., 10:815–38.

Shehabuddin, Elora. 2008. *Reshaping the Holy: Democracy, Development, and Muslim Women in Bangladesh*. New York: Columbia University Press.

Shulman, David. 2012. *More Than Real: A History of the Imagination in South India*. Cambridge, MA: Harvard University Press.

Siddiqui, Kamal, ed. 2005. *Local Government in Bangladesh*. 3rd ed. Dhaka: University Press.

Siegel, James T. 2011. "The Hypnotist." In *Objects and Objections of Ethnography*, 97–115. New York: Fordham University Press.

Singh, Bhrigupati. 2015. *Poverty and the Quest for Life: Spiritual and Material Striving in Rural India*. Chicago: University of Chicago Press.

Sirdar, M. Ansar Uddin. 1999. *Land Laws and Land Administration Manual*. Dhaka: Selima Anaju Ara.

Soeftestad, Lars. 2000. "Riparian Right and Colonial Might in the *Haors* Basin of Bangladesh." Paper presented at the conference of the International Association for the Study of Common Property, in the Panel "Constituting the Riparian Commons," Bloomington, IN, May 31–June 4.

Srinivas, M. N., and A. M. Shah. 1960. "The Myth of Self-Sufficiency of the Indian Village." *Economic Weekly*, September 10, 1375–78.

Steinby, Lisa. 2009. "The Rehabilitation of Myth: Enlightenment and Romanticism in Johann Gottfried Herder's Vom Geist der Ebräischen Poesie." *Sjuttonhundratal* 6:54–79.

Steward, Julian H. (1955) 1972. *Theory of Culture Change: The Methodology of Multilinear Evolution*. Urbana: University of Illinois Press.

Stewart, Tony K. 2001. "In Search of Equivalence: Conceiving Muslim-Hindu Encounter through Translation Theory." *History of Religions* 40 (3): 260–87.

Stewart, Tony K. 2012. *The Final Word: The Caitanya Caritamrita and the Grammar of Religious Tradition*. Oxford: Oxford University Press.

Strathern, Marilyn. 2006. "Divided Origins and the Arithmetic of Ownership." In *Accelerating Possession: Global Futures of Property and Personhood*, edited by Bill Maurer and Gabriele Schwab, 135–73. New York: Columbia University Press.

Strathern, Marilyn. 2009. "Land: Intangible or Tangible Property?" In *Land Rights*, edited by Timothy Chesters, 13–38. Oxford: Oxford University Press.

Thorp, John P. 1978. *Power among the Farmers of Daripalla: A Bangladesh Village Study*. Dhaka: Caritas.

Thorp, John P. 1982. "The Muslim Farmers of Bangladesh and Allah's Creation of the World." *Asian Folklore Studies* 41 (2): 201–15.

Tinker, Hugh. 1968. *The Foundations of Local Self-Government in India, Pakistan and Burma*. New York: Frederick A. Praeger.

Trawick, Margaret. 1992. *Notes on Love in a Tamil Family*. Berkeley: University of California Press.

Tsing, Anna Lowenhaupt. 2015. *The Mushroom at the End of the World: On the Possibility of Life in Capitalist Ruins*. Princeton, NJ: Princeton University Press.

van Schendel, Willem. 1981. *Peasant Mobility: The Odds of Life in Rural Bangladesh*. Amsterdam: Van Gorcum.

van Schendel, Willem. 2009. *A History of Bangladesh*. New York: Cambridge University Press.

Webster, Peter J. 1987. "The Elementary Monsoon." In *Monsoons*, edited by Jay S. Fein and Pamela L. Stephens, 3–32. New York: John Wiley and Sons.

Weinstein, Bernard. 2017. "Liberalism, Local Government Reform, and Political Education in Great Britain and British India, 1880–1886." *Historical Journal* 1: 1–23.

Westergaard, Kirsten. 1985. *State and Rural Society in Bangladesh: A Study in Relationship*. London: Curzon Press.

Westergaard, Kirsten, and Muhamad Mustafa Alam. 1995. "Local Government in Bangladesh: Past Experiences and Yet Another Try." *World Development* 23 (4): 679–90.

Westergaard, Kirsten, and Abul Hossain. 2005. *Boringram Revisited: Persistent Power Structure and Agricultural Growth in a Bangladesh Village*. Dhaka: University Press.

White, Leslie. (1959) 2016. *The Evolution of Culture: The Development of Civilization to the Fall of Rome*. London: Routledge.

White, Sarah. 1992. *Arguing with the Crocodile: Gender and Class in Bangladesh*. Chicago: Zed Books.

Wilce, James M. 1998. *Eloquence in Trouble: The Poetics and Politics of Complaint in Rural Bangladesh*. New York: Oxford University Press.

Wirth, Jason M. 2007. "Foreword." In Friedrich Wilhelm Joseph von Schelling, *Historical-critical Introduction to the Philosophy of Mythology*, vii–xiv. Albany: State University of New York Press.

Wright, David K., J. Andrew Darling, Barnaby V. Lewis, Craig M. Fertelmes, Chris Loendorf, Leroy Williams, and M. Kyle Woodson. 2013. "The Anthropology of Dust: Community Responses to Wind-Blown Sediments within the Middle Gila River Valley, Arizona." *Human Ecology* 41 (3): 423–35.

Yanagisako, Sylvia Junko. 2015. "Kinship: Still at the Core." HAU: *Journal of Ethnographic Theory* 5 (1): 489–94.

Zafarullah, Habib, and Ahmed Shafiqul Huque. 2001. "Public Management for Good Governance: Reforms, Regimes, and Reality in Bangladesh." *International Journal of Public Administration* 24 (12): 1379–403.

Zaidi, S. M. Hafeez. 1970. *The Village Culture in Transition: A Study in East Pakistan Rural Society.* Honolulu: East-West Center Press.

Zakaria, Saymon. 2011. *Pronomohi Bongomata: Indigenous Cultural Forms of Bangladesh.* Dhaka: Nymphea Publications.

Zaman, M. Q. 1996. "Development and Displacement in Bangladesh: Toward a Resettlement Policy." *Asian Survey* 36 (7): 691–703.

borsha (rainy season), 62
Brammer, Hugh, 207n9
Brandel, Andrew, 18
Breman, Jan, 10
Burn, Sir Richard, 139, 144
Butler, Judith, 84

Cassirer, Ernst, 162, 211n2, 211n3
Chasin, David, 166
Chaitanya, 147
char (newly deposited lands), 21, 25: cultivation of, 29, 87, 207n7; and development projects, 21; emergence of, 29–30, 73–74, 76, 82, 131–32; legality of, 7, 19, 36; as property, 28–58
char dweller. See chaura
chashi (farmer), 110. See also agriculture
Chatterji, Joya, 143
chaura (char dweller): discrimination against, 2; elections of, 94–130; form of life, 4, 9, 11, 64, 194; government neglect of, 5, 11; hierarchy of, 8, 179; labor of, 8, 29, 52, 56, 132, 203n19; movement of, 7, 59, 82, 84–85, 88, 96, 120, 156, 175, 178, 181; self-opacity of, 7 9–85, 209n1; sociality of, 6–7, 19, 30–31, 55, 88, 95, 138–40, 149–50; women, 8, 20, 52–55, 65–69, 79, 82–85, 106, 110, 160–64, 174, 177–85, 203n19, 203n20
children: death of, 160–90; lives of, 175–77
Chitra Nodir Pare (dir. Mokammel), 151–52
Chittick, William, 194
Choudhury, Nurul H., 141
Chowdhury, Jodunath, 152–53
Chowdhury, Mohini, 185, 195
climate change, 3, 176, 194, 197n2, 206n4
Collingwood, R. G., 9
colonialism, 4–7, 30–31, 35–37, 95–96, 104–8, 138–44, 170, 193, 199n4
Cotton, James Sutherland, 139
culture: and nature, 9, 13, 18, 163, 181, 185; reproduction of, 10, 162, 181

dag (plot number), 38, 45, 49
dalil (deed of ownership), 30–31, 38–40, 48–50, 52–53, 148–49, 153
Darian, Steven, 165, 173–74
Das, Veena, 18, 161, 210n1

death: of children; 160–90
debt, 33, 68
democracy, 74, 95, 105, 151. See also electoral politics
Descartes, Rene, 204n21
Descola, Philippe, 9
Dill-Riaz, Shaheen, 163, 169, 181, 212n11
din mojur (day laborer), 110
displacement: aftermath of, 131–32, 145; anthropocentric, 15; of chauras, 19, 59, 71, 85, 87, 109–10; of Hindus, 193; narratives of, 178, 193
dispossession, 138, 145–46, 193
D'Souza, Rohan, 207n11
Dumézil, Georges, 212n7
Durkheim, Émile, 198n2

East Bengal State Acquisition and Tenancy Act (1950), 36, 201n14
Eaton, Richard, 171
electoral politics: 17, 94–131; and local self-government, 104–16; narrativization of, 126; national, 97–98; and natural landscape, 119–21; and performativity, 117; in Sirajganj, 98–104; and virtual villages, 116–19
embodiment, 9, 14, 42, 193, 212n11
Enemy Property Act (1968), 143, 145
Engels, Friedrich, 29
entrainment, 60–61, 77–79, 89
erosion: and entrainment, 77–79; experience of, 60, 74–77, 87–89, 98, 119, 126, 209n1; narratives of, 19, 60, 80, 193
Ershad, Muhammad, 71, 207n9, 207n10
estrangement, 151–56
ethics, 162. See also morality
everyday, the: and erosion, 77–79; and myth, 161–62, 171, 179, 181; and the nation-state, 59–62
evil, 17–19, 136–38, 156, 161, 168, 183, 209n1

family. See poribar
famine, 11, 60–61, 72–73, 207n10
Faraizi movement, 7, 141
Feldman, Shelley, 145
Fichte, Johann Gottlieb, 15
film, 118, 138, 151–53, 155–56, 163, 169, 181. See also Sand and Water (dir.

Dill-Riaz); *Titash Ekti Nodir Naam*
(dir. Ghatak)
flood. *See bonna*
Flood Action Plan, 71, 207n9
forgetting, 17, 60, 84, 135–38, 151, 177, 179–80
freedom, 16, 61, 137, 174, 194

Geisler, Charles, 145
geo-ontology, 14, 193
ghost story, 138, 151, 155–56
Gilmartin, David, 98
Gluckman, Max, 198n2
Goethe, Johann Wolfgang von, 17, 96
gondogol (time of difficulty), 50
Gongima (Ganga Devi), 20, 161–62, 164–67, 170–76, 179–85
grihosto (householder), 48, 110
Guha, Ranajit, 35
Guhathakurta, Meghna, 145
gusti (clan), 46, 53, 83, 100, 117

Haraway, Donna, 13
Hadot, Pierre, 9
Harris, Michael, 203n18
Harrison, Simon, 135
Hasina, Sheikh, 97
hath (unit of measurement), 63, 75
Hegel, Georg Wilhelm Friedrich, 16, 29, 135
Hinduism, 142, 147, 166, 170–71
Hindus: active absence of, 138, 143–46; exodus of, 143–44; contemporary perspectives of, 146–50; and Muslim relations, 151, 156, 162
Hofer, Thomas, 204n1, 206n5
hospitality, 60–61, 132, 143
Hossain, Naomi, 11
Hunter, Sir William Wilson, 139

imagination, 9, 19, 88, 137, 154, 162, 194; and electoral politics, 96; Faraizi, 7; national, 74, 204n1; and nature, 17, 19, 96, 126. *See also* mind; myth
India-Pakistan War (1965), 201n13
Indigo Rebellion (1859), 141
inheritance: 52–54, 56, 86, 137, 168, 199n5, 202n18, 203n20; law, 52, 202n15; loss of, 54; renunciation of, 54

injustice, 18, 138, 156, 168, 177
intuition, 17, 168, 194
Iqbal, Muhammad, 174
Islam, 7, 56, 143, 147, 161, 167, 170–72, 185
Islam, A. K. M. Aminul, 117, 208n4
Islamization, 151, 185, 210n4

Jabbar, Abdul, 154
Jalais, Annu, 212n5
Jamous, Raymond, 211n3
Jamuna River: complexity of, 125; course of, 78, 139; geography of, 1; as system, 3, 96, 119, 124–26
jasus (spy), 102
jolaha (weaver), 133, 140

Kali, 136, 140, 142, 153, 165, 212n5
Kant, Immanuel, 15–17, 204n21
Karim, A. H. M. Zehadul, 200n6
kayija (fight), 34–35, 40, 43–46, 52–54
khajna (rent), 64, 110, 133
Khan, Mohammed Ayub, 105, 107, 151
Khan, Mushtaq, 97
khas (government-owned land), 35–37, 39, 45–46, 50–51, 81, 135, 144–45, 178, 201n9; and land laws, 35–37
khatiyan (record of rights), 30, 38, 45, 49
Khijir, Khwaja, 20, 161–76, 180–85, 212n6
kinship, 42, 61, 85, 88, 120, 128, 199n5, 199n6, 211n13; fictive, 62; and land, 19, 32–33, 40–42, 46–47, 50, 55–56, 98, 125, 193, 202n16; matrilineal, 42, 52–53
Kloos, Peter, 108
Kohn, Eduardo, 14
kot (lease agreement), 133
Kotalova, Jitka, 177

labor: agricultural, 8, 18, 31, 41–42, 62, 64, 110, 131, 140, 199n13, 201n12, 206n4; of *bhangon* (domestic dismantling), 63, 66–68, 70, 80; domestic, 66, 182–83; emotional, 52; mental, 29, 56; physical, 29, 41; transactional, 52
land: in absentia, 6, 29, 32; accretion of, 2, 6, 11, 36, 76; ancestral, 64, 133, 148; cultivation of, 32, 29, 40–41, 48, 50, 60, 64, 87, 131, 138; and landlessness, 29, 31, 33, 59, 73, 202n18

Rahman, Atiur, 203n18
Rahman, Rezaur, 205n1
Rahman, Sheikh Mujibur, 73
raiyat (peasant), 35–36, 140, 199n4
Rappaport, Roy, 13
Rashid, Haroun Er, 75
Rashid, Salim, 205n1
Ratzel, Friedrich, 12
religion: 56, 147, 171, 212n8; Christianity, 147; Islam, 7, 56, 143, 147, 161, 166–67, 170–72, 185; Hinduism, 142, 147, 166, 170–71; Shaivism, 142, 166, 210n3; Vaishnavism, 142, 155, 166, 210n3
Riaz, Ali, 97
riot, 140–41

Sahlins, Marshall, 13, 32
Saith, Ashwani, 108
Sand and Water (dir. Dill-Riaz), 181–83
Sayidur, Mohammad, 108
Sarkar, Rashid, 147
Schelling, Friedrich Wilhelm Joseph von, 15–18, 61, 137, 194, 209n1, 211n2; *First Outline of a System of a Philosophy of Nature*, 16; *Historical-Critical Introduction to the Philosophy of Mythology*, 162; *Ideas for a Philosophy of Nature*, 18; *Philosophical Investigations into the Essence of Human Freedom*, 137; *System of Transcendental Idealism*, 16, 61
Schendel, Willem van, 106
Scheper-Hughes, Nancy, 163
Schneider, David, 199n6
Sepoy Rebellion (1857), 141
Severi, Carlo, 171, 173
Shah, A. M, 117
shalish (dispute settlement)
"Shonkhraj" (Jabbar), 154–55
shomaj (society), 48, 117–18, 209n6
Siegel, James, 209
Singh, Bhrigupati, 212n7
Srinivas, M. N., 117
Steward, Julian, 13
Stewart, Tony, 171
Strathern, Marilyn, 198n1, 198n2
suffering, 10, 70, 72, 76, 87, 178, 193, 207n8

Tenancy Act (1938), 30, 36
Tenant Revolt (1873), 141
Thorp, John, 117, 209, 213n13
Titash Ekti Nodir Naam (dir. Ghatak), 118

unconscious, the: and entrainment, 89; and nature, 17, 19, 62, 85, 126, 137, 180–81; in Schelling, 61, 194, 209n1
upazila (subdistrict), 31, 94, 104–5, 142, 146–48

Vested Property Act (1974), 135, 145
village: and local self-governance in Bangladesh, 108–9; as unit of political representation, 108;
virtuality of, 116–18
violence, 7, 35, 55, 144, 161, 209n2, 210n1; aestheticization of, 34; and electoral politics, 97, 122; and land disputes, 31–32, 34–35, 39, 53, 55, 146, 209n2, 210n1; and women, 53–54. *See also* kayija; war

war, 10, 73, 86–87; India-Pakistan, 36, 143, 153, 201n13; Liberation War of, 44, 50, 52, 72, 144, 153
Westergaard, Kirsten, 201n14, 207n10, 208n6
White, Leslie, 13
Wilce, James, 178
Wilson, E. O., 13
women: as assets, 85; and *bhangon*, 66–69; and electoral politics, 106, 110, 119, 128; and loss, 20, 160–64, 179–80, 183, 185; memory of, 80, 82–83, 180, 183; and property, 52–55, 203n20; speech of, 160–64, 177–81; status of, 203n19, 209n6
Wright, David K., 77

Zaidi, Hafeez, 212n10
Zakaria, Saymon, 200n8
zamindar (landed gentry), 4, 35, 45, 140–42, 144–45, 150, 153–55; abolishment of, 36–37, 45, 49, 201n14. *See also* property
zila (district), 36, 94, 105, 117

www.ingramcontent.com/pod-product-compliance
Lightning Source LLC
Chambersburg PA
CBHW071737270326
41928CB00013B/2706